Modern Livestock and Poultry Production

Dr. Ram Prakash Singh was borne on 30th April 1963 in Itawa. Presently he is a senior scientist in Agricultural Extension and faculty member in the division of Dairy Extension. He obtained his bachelor degree in agriculture and completed his M.Sc. and P.hD. in Dairy Extension during 1985 and 1989 respectively. He has handle two research projects and principal investigator and was associated with many prestigious projects. Presently he is engaged in research and extension in Western U.P. with special emphasis on livestock production.

Modern Livestock and Poultry Production

by

Ram Prakash Singh

BIOTECH

2023

BIOTECH BOOKS

ISBN 978-81-7622-197-9

BIOTECH BOOKS
4762-63/23, Ansari Road, Darya Ganj,
New Delhi - 110 002
Ph.: 43003222
e-mail : biotechbooks@yahoo.co.in

PRINTED IN INDIA

Preface

A large proportion of the world's land area is unsuitable for crop production because of topography, climate and deficiencies in soil fertility or rainfall. In many such areas, the use of grazing animals is the only feasible way economically to convert the natural materials of plant growth into products of use to people. This has evolved into animal husbandry, the art and science of the care, management and improvement of domestic animals and their products. The manipulation of animals for the benefit of human beings is an integral part of the story of civilisation. Animal production in different areas and cultures of the world reflects the importance of domestic animals as sources of food, power, fibre, hides and skins.

This text has been aimed at those involved in animal husbandry as a guide into the production techniques involved in livestock and poultry. The traditional curriculum for the study of animal husbandry deals with the technology of cattle, horse, pig, swine, sheep, goat and poultry production, and the book undertakes a detailed look into the breeding, rearing and feeding techniques that have a telling effect on the growth or shrinkage of livestock numbers not ignoring the effect of natural conditions. In addition, space has been given to understanding the essential prospects, challenges and changes offered by these livestock and poultry production technologies.

Ram Prakash Singh

Contents

1

Modern Trends in Livestock Production

The term "livestock" livestock refers to any breed or population of animal kept by humans for a useful, commercial purpose. This can mean domestic animals, semi-domestic animals, or captive wild animals. Semi-domesticated refers to animals which are only lightly domesticated or of disputed status. These populations may also be in the process of domestication.

Livestock' are defined, in part, by their end purpose as the production of food or fiber, or labour. The economic value of livestock includes:

— *Meat*: the production of a useful form of dietary protein and energy.

— *Dairy products*: Mammalian livestock can be used as a source of milk, which can in turn easily be processed into other dairy products such as yogurt, cheese, butter, ice cream, kefir, and kumis. Using livestock for this purpose can often yield several times the food energy of slaughtering the animal outright.

— *Fiber*: Livestock produce a range of fiber/textiles. For example, sheep and goats produce wool and mohair; cows, deer, and sheep can make leather; and bones, hooves and horns of livestock can be used.

— *Fertiliser*: Manure can be spread on fields to increase crop yields. This is an important reason why historically, plant and animal domestication have been intimately linked. Manure is also used to make plaster for walls and floors and can be used as a fuel for fires. The blood and bone of animals are also used as fertiliser.

- *Labour*: Animals such as horses, donkey, and yaks can be used for mechanical energy. Prior to steam power livestock were the only available source of non-human labour. They are still used for this purpose in many places of the world, including ploughing fields, transporting goods, and military functions.
- *Land management*: The grazing of livestock is sometimes used as a way to control weeds and undergrowth. For example, in areas prone to wild fires, goats and sheep are set to graze on dry scrub which removes combustible material and reduces the risk of fires.

During the history of animal husbandry many secondary products have arisen in an attempt to increase carcass utilisation and reduce waste. For example, animal offal and non-edible parts may be transformed into products such as pet food and fertiliser. In the past such waste products were sometimes also fed to livestock as well. However, intra-species recycling poses a disease risk, threatening animal and even human health. Due primarily to BSE (mad cow disease), feeding animal scraps to animals has been banned in many countries, at least in regards to ruminants.

Livestock are generally kept in an enclosure, are fed by human-provided food and are intentionally bred, but some livestock are not enclosed, or are fed by access to natural foods, or are allowed to breed freely, or any combination thereof.

Livestock raising historically was part of a nomadic or pastoral form of material culture. The herding of camels and reindeer in some parts of the world remain unassociated with sedentary agriculture. The trans-humance form of herding in the Sierra Nevada Mountains of California still continues as cattle, sheep or goats are moved from winter pasture in lower lying valleys to spring pasture and summer pasture in the foothills and alpine regions as the seasons progress. Cattle were raised on the open range in the Western United States and Canada, as well as on the Pampas of Argentina and other prairie and steppe regions of the world.

The enclosure of livestock in pastures and barns is a relatively new development in the history of agriculture. When cattle are enclosed, the type of 'enclosure' may vary from a small crate or to a large fenced pasture. The type of feed may vary from natural growing grass, to highly sophisticated processed feed. Animals are usually intentionally bred through artificial insemination or through supervised mating.

Indoor production systems are generally used only for pigs and poultry, as well as for veal cattle. Indoor animals are generally farmed intensively, as large space requirements would make indoor farming unprofitable and impossible. However, indoor farming systems are controversial due to: the waste they produce, odour problems, the potential for groundwater contamination and animal welfare concerns.

Other livestock are farmed outside, although the size of enclosure and level of supervision may vary. In large open ranges animals may be only occasionally collected in "round-ups" or "musters". Herding dogs such as sheep dogs and cattle dogs may be used for mustering as are cowboys, stockmen and jackaroos on horseback or in helicopters. Since the advent of barbed wire (in the 1870s) and electric fence technology, fencing pastures has become much more feasible and pasture management simplified. Rotation of pasturage is a modern technique for improving nutrition and health while avoiding environmental damage to the land. In some cases very large numbers of animals may be kept in indoor or outdoor feeding operations (on feedlots), where the animals' feed is processed, offsite or onsite, and stored onsite then fed to the animals.

Livestock—especially cattle—may be branded to indicate ownership, but in modern farming identification is more likely to be indicated by means of ear tags than branding. This is not only more humane, but also has other advantages such as reducing the likelihood of infection and damage to the livestock. Sheep are also frequently marked by means of ear tags. As fears of mad cow disease and other epidemic illnesses mount, the use of microchip identification to monitor and trace animals in the food production system is increasingly common, and sometimes required by governmental regulations.

Modern farming techniques seek to minimise human involvement, increase yield, and improve animal health. Economics, quality and consumer safety all play a role in how animals are raised. Drug use and feed supplements (or even feed type) may be regulated, or prohibited, to ensure yield is not increased at the expense of consumer health, safety or animal welfare.

Drivers in the Livestock Sector

Individual consumption of livestock products is closely related to per capita income. That is, with growing incomes people typically increase their consumption of meat, milk and eggs until these products become fully integrated into the daily diet. In high-income countries, per capita consumption of meat ranges between 80 kg and 130 kg per year; there is practically no further increase beyond that level. As incomes in many developing countries have grown rapidly over the past 20 years, consumption levels of meat and other livestock products have also increased. The economies of developing countries' achieved an average annual growth of 3.8% (1.8% per capita) from 1991 to 2001, up from 2.9% during the ten preceding years.

Developing countries in East Asia, in particular, have experienced very strong economic growth, with an annual rate of 7.4% (6.2% per capita) over the decade between 1991 and 2001, with the People's Republic of China leading as the world's fastest growing economy. South Asia and the Near East follow, with gross domestic product growth rates of 5.5%

and 4.4% over the same period. Economic expansion has been more modest in Latin America, at 2.9% annually, and in sub-Saharan Africa at 2.6%.

In addition to higher incomes, increases in human populations add to the demand for animal-source food products. Most developing countries still have rapidly growing populations even though percentage growth rates are below their peak in the 1970s. Each year, the human These adjustments are based on a changing feed resource base, particularly feed concentrates. Current and projected levels of livestock production would not be possible without the expanding production and yield increase of crop agriculture.

Traditionally, livestock production used to be based on locally available feed resources, including local fodder, crop residues and unconsumed parts of human food—resources that had no value as human food. Traditionally, natural pastures were the venue of livestock production. More recently, however, a growing proportion of pasture land in developing countries is in areas which are unfit or marginal for cropping, and degraded arable land is often converted into pasture land. The demand for arable land, and the fact that there is basically no additional available land that can readily be converted into pastures, except in parts of tropical Latin America, have important implications for the livestock sector. The lack of available new land prohibits a 'horizontal' expansion of existing modes of production, and forces the sector into rapid technological change and search for alternative resources.

Livestock Production Systems

Production environments, and the intensities and purposes of production, vary greatly within and across countries. Animal agriculture systems have been categorised on the basis of agro-ecological opportunities and demand for livestock commodities. In general, these systems are shaped by prevailing biophysical and socio-cultural environments, and without external inputs they have traditionally been mostly in sustainable equilibrium with such environments. In many of these systems, the livestock element is interwoven with crop production, as in the rice/buffalo or cereal/cattle systems of Asia.

Animal manure is often essential for maintaining soil fertility, and the role of animals in nutrient cycling is often an important motivation for keeping animals, particularly where this involves a transfer of nutrients from common property resources to private land. In other cases, such as the semi-nomadic pastoral systems of the world's natural grassland regions, environmentally stable balances of human society, animal population and vegetative biomass have been maintained for centuries. Many of these systems that are the result of a long evolution are currently under pressure to adjust to rapidly evolving socio-economic conditions; large intensive livestock production units, in particular for pig and poultry production, have emerged over the last decades in many developing regions in response to the rapidly growing demand for livestock products. To classify the resulting continuity of situations into a limited number of distinct LPS, the following criteria should ideally be considered:

- integration with crops
- relation to land
- agro-ecological zone
- intensity of production
- type of product.

The classification, which uses only the first three of these classification criteria, identifies ten broad categories of systems. The LPS are considered a subset of farming systems. There are two main groups of LPS: those solely based on animal production and those where cropping and livestock rearing are associated. The first group is defined as systems in which more than 90% of dry matter fed to animals comes from rangelands, pastures, annual forages and purchased feeds, and less than 10% of the total value of production comes from non-livestock farming activities.

Mixed farming systems are livestock systems in which more than 10% of the dry matter fed to animals comes from crop by-products such as stubble, or where more than 10% of the total value of production comes from non-livestock farming activities. Landless LPS are a subset of the pure livestock systems in which less than 10% of the dry matter fed to animals is farm produced and in which annual average stocking rates are above ten livestock units per hectare of agricultural land (on average at census unit level). Grassland-based systems are defined as systems in which more than 10% of the dry matter fed to animals is farm produced and in which annual average stocking rates are less than ten livestock units per hectare of agricultural land.

A distinction is made between:

— temperate zones and tropical highland

— humid/sub-humid tropics and sub-tropics

— arid/semi-arid tropics and sub-tropics.

Rainfed mixed farming systems are mixed systems in which more than 90% of the value of non-livestock farm production comes from rainfed land use; these systems can be divided into the same agro-ecological sub-classes as given above. Irrigated mixed farming systems are systems in which more than 10% of the value of non-livestock farm production comes from irrigated land use, and again includes the same sub-classes.

The presence of industrial systems is connected to both demand factors and supply determinants; areas with high population density and purchasing power, in particular coastal areas in East Asia, Europe and North America, which also have access to ocean ports, show a high prevalence of industrial systems and import much of the necessary feed. In contrast, there are areas with ample feed supplies such as the mid-western United States of America (USA) and interior parts of Brazil and Argentina, where industrial systems rely mainly on local feed surpluses.

East and Southeast Asia strongly dominate industrial monogastrics' production in the developing regions. Southern

Brazil is another industrial production hot spot at world level, while important regional centres of industrial production are found, for example in Mexico, Colombia, Venezuela and Chile. Similarly there are major regional centres for the industrial production of chicken in Nigeria, South Africa and the Middle East.

Of the land-based system group, grazing systems cover the largest area and are currently estimated to occupy some 26% of the earth's ice-free land surface. This figure obviously includes a large variety of agro-ecological contexts with very different levels of biomass productivity. Grazing systems are primarily found in the more marginal areas which are unfit for cropping because of topography, low temperature or low rainfall.

Mixed farming systems are prevalent in bio-climatically more favoured ecosystems. Most mixed farming systems are rain-fed, and they are particularly widespread in semi-arid and sub-humid areas of the tropics and in temperate zones. Mixed irrigated systems are found throughout the world, but in most cases are relatively small. Exceptions are the eastern parts of the People's Republic of China, northern India and Pakistan, where mixed irrigated systems extend over large areas.

Only a small fraction of the world's ruminant population is found in industrial feedlots, partly because this corresponds to only the final stage of the animal's life cycle, even in regions where intensive production is common. Ruminant feedlots are predominantly a North American phenomenon, though they are used to a lesser extent in parts of Europe and the Near East. The vast majority of large and small ruminant populations are found in the developing regions: some 70% of small ruminants in grazing systems and over 80% of large ruminants in grazing systems are located in developing regions. These shares are respectively about 80% and 70% in rainfed mixed systems and 87% and 92% in irrigated mixed systems.

Ruminant productivity varies considerably within each system, but in grazing and mixed systems overall productivity is lower in developing countries than in developed ones: in grazing systems, for example, worldwide beef production per head averages 36 kg/head/year, but the average for developing countries is only 29 kg/head/year. In the mixed rainfed system, the difference between developed and developing regions is even more marked. By far the largest variation in intensity of production is found within this system, which is the largest producer of ruminant products. Even though the developing regions host the vast majority of the mixed rainfed ruminant population, they account for less than half of the system's production worldwide.

In fact beef productivity in these regions averages 26 kg/head, as opposed to 46 kg/head at world level, and their milk production represents only 22% of the world total. Across all systems, developing regions account for half of the world's beef production, some 70% of mutton production and about 40% of milk production. A sharply contrasting situation is found in the monogastric sector. More than half of the world's pork production currently originates from industrial systems, and over 70% of poultry meat. About half of this production originates from developing countries and, though reliable population figures are not available, variation in productivity between regions is probably much lower than for ruminants.

Including the substantial monogastric production from irrigated mixed systems in developing regions, these regions account for the majority of the world's pork, poultry and egg production. Huge differences are found between the developing regions: although substantial, total production in Latin America is less than one tenth of that in Asia, and production in Africa and the Near East is almost non-existent. The developed countries and Asia together account for over 95% of the world's industrial pork production.

There has recently been a rapid growth in industrial production of monogastrics in the tropics and sub-tropics, leading to production levels that are similar to those of temperate regions. However, the situation is very different for ruminant production, partly because of its land-based nature;

production and productivity are much higher in the cooler climates. Small ruminant production in the (semi-)arid (sub-)tropics is a notable exception, explained by the large population and the relatively high rates of breeding, the latter being due to the fitness of these species, which are well adapted to harsh and marginal conditions. The relatively low productivity for milk in the more humid tropics relates to the strong dominance of mixed systems in these regions, where substantial use is still made of animals for draught power and other nonproductive uses.

Growth of Livestock Production

Intensification of livestock production is taking place with regard to the use of most of the production inputs. In particular, the intensity of feed use has greatly increased over recent decades. While a growing number of people in the developing world are moving up the food chain, enjoying a richer and more diverse diet, so too are livestock; traditional fibrous and energy-rich feed stuffs are in relative decline, and protein-rich feeds together with sophisticated additives that enhance feed conversion are on the rise.

On balance, pasture land productivity has lagged far behind that of cultivated areas, although detailed estimates are difficult to make. A number of factors contribute to this trend. First, making more intensive use of the areas classified as pastures is often technically difficult and unprofitable. Constraints on the productivity of pastures most commonly relate to climatic features, topography, shallowness and/or acidity of the soils, and disease pressure, among other factors. These constraints can be overcome only with massive investments to address them on various fronts. Additionally, in much of Africa and Asia, most pastures are used as common property, which further complicates attempts to intensify production.

Without firm institutional arrangements, private investments in these areas are difficult to organise as returns accrue to individuals in proportion to the number of livestock they graze on communal land. Limited market access due to lack of infrastructure in these remote areas further contributes to the difficulty in achieving productivity improvements through individual investments. The harsh conditions of these pasturelands are exemplified by the pastoralist and agro-pastoralist areas in arid and semi-arid lands in sub-Saharan Africa.

As livestock production grows and intensifies, it depends less and less on locally available feed resources but increasingly on feed concentrates that are traded domestically and internationally. In 2004, a total of 690 million tonnes of cereals were fed to livestock (34% of the global cereal harvest) and another 18 million tonnes of oilseeds (mainly soya). In addition, 295 million tonnes of protein-rich processing by-products were used as feed (mainly bran, oilcakes and fish meal). Species that can profitably make use of such feed concentrates (pigs and poultry) have an advantage over those that cannot do this to the same extent (cattle, sheep, goats). Among the monogastrics, it is poultry that shows the highest growth rates and lowest costs per unit of output, mainly because of efficient feed conversion. The use of feed concentrates for ruminants is limited to countries with low grain-meat price ratios. Where these ratios are high, typically in grain- or cereal-deficit developing countries, grain feeding to ruminants is usually not profitable.

What is driving the increasing use of feed grains? First, there is a decline in grain prices, a trend that is basically unchanged since the 1950s—which implies that global agriculture has been in a position to easily meet the growing demand over that period. The total supply of cereals increased by 46% over the 24 years from 1980 to 2004. In real terms (constant US dollars), international prices for grains are now half those in 1961. Expanding supply at declining prices has been brought about by intensification of existing crop-land use (not by expanding the crop area: globally, cereal crop-land declined by 5.2% between 1980 and 2004).

Intensification is a result of technological advances and greater use of inputs in crop production, notably in the areas of plant breeding, irrigation and water management, application of fertilisers and mechanisation. In contrast to developed countries, the expansion of the area dedicated to cereals has been an important contributor to increasing supplies in the developing countries, with rates highest in sub-Saharan Africa (64%) and East and Southeast Asia (15.2%) over the period 1980 to 2004. However, the vast majority of the additional supplies in these regions are used for food and not for feed.

In Latin America, expansion of the area dedicated to cereal production has been slower (3.9%), but there was a 97% increase in the area under oil crops, notably soya. Here, however, the demand for feed crops has been a major determining factor in the expansion of arable land. Some countries have seen a particularly strong expansion of cropped area, most of it at the expense of the forests (Brazil and other Latin America countries). Large parts of this expansion are devoted to production of feed concentrates, notably soya and maize. Intensification also draws on technical improvements in livestock production, such as genetics, health and farm management, which have contributed to raising resource use efficiency and higher output per animal.

Over the 24 years from 1980 to 2004, the pork, chicken and milk offtake per unit of stock increased by 61%, 32% and 21% respectively. Such biological and technical advances must be adapted to local conditions if they are to be profitably introduced. These advances are supported by increasing use of external services and by the specialisation of production, with a substantial shift from backyard and mixed systems to commercial, specialised, single-product operations.

Geographic Concentration

Driven by growing demand and market opportunities, and supported by technological change, the distribution of production is no longer determined by the agro-ecological potential of a given location but by a variety of interacting factors. As well as local supply of feed and demand for livestock products, driving factors include transport costs, disease concerns, environmental regulations and a whole set of other policy factors.

As countries industrialise, they follow a pattern in relocating livestock production. Livestock production has traditionally been based on locally available feed resources, particularly those that have no other use or are of limited value, such as natural pasture and crop residues. In pre-industrialised contexts, the distribution of ruminant livestock can be explained by the availability of such resources, while the distribution of pigs and poultry closely matches that of humans, because of these animals' role as waste converters.

For example, in Vietnam, a country that can be considered to be in the early stages of industrialisation, in 1991 90% of the poultry distribution pattern could be explained by the distribution of the human population. As soon as urbanisation and economic growth translate rising incomes into 'bulk' demand for animal food products, large-scale operators emerge that are initially located close to towns and cities. Livestock products are among the most perishable products, and their conservation without chilling or processing poses serious problems. Therefore, food items from livestock have to be produced in the vicinity of demand—with concomitant human health and environmental problems stemming from the rapid urbanisation of both humans and animals in the same places.

In a subsequent phase, infrastructure and technology develop sufficiently to enable the production of livestock farther away from human populations, and livestock production shifts further away from demand centres, driven by a series of factors such as lower land and labour prices, access to feed, lower environmental standards in rural areas, tax incentives, and fewer disease problems.

There is rapid change also in the way value or food chains are organised in the livestock sector. Vertical integration provides economies of scope, ensures reliability of supply, and facilitates quality management and homogeneity of products. Importantly, it also allows producers to mitigate the wide host of pathogen threats to which the livestock sector and associated food chains are subject. At the same time, and in line with previous observations on dietary changes, vertical integration accompanies the global sourcing of both livestock production inputs and outputs, and the important moves in trade liberalisation that have occurred over the last decade.

In many developing countries, particularly in rapidly developing countries in Asia and Latin America, the continued rise in per capita incomes together with the increasing urbanisation of populations has led to the westernisation of diets and transformation of food systems. The increasing affluence of urban consumers in developing countries is associated with an increase in the predominance and reach of large-scale retail stores, in particular supermarkets, responding to and also shaping the increasing demand for convenience, variety and quality assurance by individuals and households with more disposable income and increased opportunity-costs of time.

While the main customer of the supermarkets in developing countries is the urban middle class, the competition among rival chains brings the prices of similar products down, thus also accommodating lower-income households seeking to stretch the purchasing power of their food budgets. The rapid expansion in supermarket penetration in developing countries is a fairly recent phenomenon, notable only over the last five to ten years, that is proceeding at different rates and depths in the various regions of the developing world.

The emergence of supermarkets in developing countries reflects a structural change in the way that meat, dairy products and eggs are collected, inspected, processed, packaged and supplied to consumers. It is a change that has deep impacts on livestock producers, particularly in determining who can and who cannot participate in the mainstream supply chains. A segmentation of markets can be observed, between the rapidly growing formal and the stagnating or declining informal supply chains, and between the 'wet' markets for fresh meat and the supermarket outlets of processed, frozen, packaged and branded meat. The relative significance of each market segment is tied to the level of economic development. It is closely linked to the purchasing power of households and individuals, their demand for leisure, their preferences with respect to the form and texture of meat upon purchase, and the relative value they give to notions of food that are considered 'safe'.

At the same time large-scale retailers compete in delivering consistent product quality that is demanded by their main market. The concept of 'quality' from the producers' perspective is complex, and its attributes evolve over time. The definition of quality varies according to suppliers' strategies on the one hand, and to cultural influences on the other. It includes food safety, nutrition and attributes related to the commercial differentiation of the products. Large retailers require a reliable supply of agricultural products from their suppliers (producers) with consistency in volume and in quality.

Vertical coordination presents the opportunity to keep control of operating and transaction costs while at the same time meeting high standards of food safety. It demands organisational and institutional changes in the relationship between the primary producer and the agri-food processor or supermarket distributor, giving rise either to various forms of vertically coordinated transactions (the retailer contracts suppliers and/or processors), or in the extreme form, to fully integrated systems (all units in the food chain owned by one company). Large retailers in developing countries are increasingly tending towards vertical coordination, although vertically coordinated chains may interact with informal markets that supply inputs of live animals or products.

Growth of Average Scale Production

Throughout developed and rapidly developing countries, there is a continuing tendency for production scales to grow. This is particularly pronounced for the production of monogastrics, and less so for ruminants. In contrast, certain areas with low economic growth, like parts of South Asia and sub-Saharan Africa, have not yet experienced an increase in average scale of production. The process of ever-growing scales is triggered by economies of scale: average production costs decline with an expanding scale of operations at various stages of the production process. As a result, the number of producers rapidly diminishes even though the sector as a whole may expand.

In emerging economies, the average size of operations is rapidly increasing and the numbers of livestock producers are in sharp decline. For chicken in Brazil, between 1985 and 1996, most of the growth occurred in the larger farms of the central western part of the country: farms with more than 10,000 head increased their proportion of the total population from 42% to 78%.

Similarly in Thailand, only the largest category of pig farms grew in number. However, in the Southern Luzon region of the Philippines, one of the main pig producing regions of the country, the pig numbers in commercial farms exhibited phenomenal growth between 1980 and 2000, while the number of pigs produced by smallholders as a whole also increased. Smallholders can stay in business by providing their labour input to their own farms at below market price, which works well in countries where there are limited employment opportunities in other sectors. But as soon as employment opportunities in other sectors rise, many smallholder producers opt out of livestock production.

Different commodities and different stages in the production process reveal different potential for economies of scale. This potential tends to be high in the post-harvest sectors (e.g. slaughterhouses, dairy plants). In commodity production, poultry is most easily mechanised and shows a trend towards industrial forms of production even in the least-developed countries.

Growth Rates of Livestock Production

The Livestock, Environment and Development Initiative has recently conducted an update of the world livestock production systems study by Seré and Steinfeld. The ten-year period between the two quantifications (which took averages of the periods 1991 to 1993 and 2001 to 2003) allows for a number of observations that among others confirm the importance of the above processes. Significant changes in resource endowment have brought about changes in the nature and extent of certain production systems.

Permanent pastures show an increase at world level of about 3% in that period, particularly in Central/South America and western Asia/North Africa, while slight drops are observed in sub-Saharan Africa and in the countries of the Organisation for Economic Co-operation and Development (OECD). Total arable land increased by about 5%, with almost all of this growth occurring in the developing regions. Major increases are observed in tropical highland grazing systems as well as in humid mixed irrigated systems. The irrigated area expanded by some 20% in ten years. This expansion has been particularly large in the arid and tropical highland grazing systems, as well as in the arid mixed rainfed systems.

As a consequence of the agricultural expansion, global cattle stocks are slightly up (5%), with considerable increases in stock numbers for sub-Saharan Africa, Asia and Latin America. A sharp drop (almost 50%) in animal numbers occurred in Eastern Europe and the Commonwealth of Independent States (CIS, the former Soviet Union) following the collapse of the Soviet Union and political change in other countries. World output rose by about 10% in the period of observation, with very sharp differences at regional level; Asia almost doubled cattle meat output, sub-Saharan Africa increased by 30%, Latin America by 40%, and West Asia and North Africa (WANA), albeit from a lower absolute level, by about 20%. The strongest cattle output increases occurred in the mixed systems in the humid zones.

Total meat production from small ruminants increased by about 10%, while the overall stock numbers for small ruminants remained fairly constant for the two reference periods. There have been inter-regional shifts in distribution: stock numbers increased considerably in sub-Saharan Africa and Asia, and declined sharply in Latin America and the OECD, and in particular in Eastern Europe and the CIS. The increases occurred mainly in the mixed humid systems.

The changes in monogastric animal production are more notable. Total pig meat output rose by 30% at the world level, and this increase is accounted for almost entirely by the increased output in Asia. Most regional groupings show increases in pig meat production, while for Eastern Europe and the CIS there is a drop of about 30%. Industrial pig meat production grew at a pace of about 3% per annum. Major increases also occurred in the humid and temperate mixed irrigated systems.

The total production of poultry meat grew by about 75%, the greatest expansion of all livestock products. Regional differences are pronounced, with an extremely strong expansion in Asia (about 150% of added production, representing 9% yearly growth). The growth rates are generally positive, ranging between 2% and 10% across regions, the majority originating from the expansion of industrial systems. The global production of table eggs grew by about 40%. Asia more than doubled its egg production in the period and attained a share of about 50% of world production. The landless system grew by about 4% per year.

Food Safety Management

On a global average for 1997, animal products provided about 16% of the calories in the diet, representing around 25% in developed countries and around 12% in developing countries. For proteins over the same period the proportion was about 37%, being about 56% in developed countries and 29% in developing countries. Meat and other animal products also provide essential fatty acids, vitamins and minerals. The iron in meat and meat products is easily assimilated by humans and is a key to preventing iron deficiency anaemia, a high prevalence of which has been reported in, for example, Eastern Europe. The livestock industry therefore has great global economic and nutritional significance. While the global importance of animal food increases, the related changes in the livestock sector affect the security of the food provided in many ways.

The production/demand trends described above have led to an increase in white meat consumption and a decline in red meat. This has important implications because it is the poultry, pig (requiring protein supplementation) and dairy industries that are the principal users of processed animal feeds. This feed must be of high quality to sustain efficient growth and feed conversion. It is important to realise, however, that the large and increasing volume of international trade in foods of animal origin as well as in feedstuffs adds an important international dimension to the control of animal feedstuffs. Given the direct links between feed safety and the safety of foods of animal origin, it is essential that feed production and manufacture be considered as an integral part of the food production chain.

Feed production must therefore be subject, in the same way as food production, to quality assurance, including food safety systems based preferably on hazard analysis and critical control points. The Codex Alimentarius Commission has negotiated a Code of Conduct on Safe Animal Feeding, adopted in 2004, to help meet such requirements. While in many developing countries LPS are subject to strong external forces and are in the process of rapid adjustment, little change is occurring in other countries.

In a number of ways, it is the dichotomy that characterises livestock production in developing countries that poses the greatest challenge to food safety. Where traditional and modern forms of livestock production co-exist, with parallel market channels and outlets, uniform standards are difficult to enforce because of equity concerns, disease concerns, certification problems and other issues. This poses a formidable challenge to regulating and upgrading food safety in developing countries, particularly also because the two segments interact.

Food safety requirements are a major determinant shaping the structure of the livestock sector and associated food chains.

One of the main aspects of the structural change process described above is the de-coupling of production locations from consumption, and this has resulted in the risk through the food chain becoming an issue. Vertical integration, one of the other characteristics of structural change, also may affect food safety. The emergence of multinational food chains and a dramatic rise in the market share of supermarkets and the modern retail sector in many countries have created a number of effects, some of which may affect food safety:

- — a shift toward cross-border systems, with corporations procuring goods in their different countries of operation—a shift toward preferred-supplier systems to select producers who meet specific quality and safety standards and lower transaction costs
- — a consolidation of production and processing which may have unexpected impacts on safety
- — a shift toward safety and quality standards driven by the private sector.

References

FAO, *Expert Consultation on Application of effective herd health and production programmes to increase livestock productivity in developing countries,* Rome, 30 March-1 April 1993.

_______________ , "World livestock production systems - current status, issues and trends", *FAO Animal Production and Health Paper,* No. 127. Rome, 1996.

Holden, S., Ashley, S. & Bazeley, P., *Improving the delivery of animal health services in developing countries: a literature review,* Somerset, UK, Livestock in Development, 1996.

McCorkle, C.M. & Mathias, E., "Paraveterinary healthcare programmes: a global overview", *In* K.-H. Zessin, ed. *Livestock production and diseases in the tropics: livestock production and human welfare,* Proceedings of the VII International Conference of Institutions of Tropical Veterinary Medicine, Vol. II, p. 544-549, Berlin, 25-29 September 1994.

Umali, D., Feder, G. & de Haan, C., "The balance between public and private sector activities in the delivery of livestock services", *World Bank Discussion Papers,* No. 63. Washington, DC, World Bank, 1992.

2

Livestock Production Systems

Livestock production systems are agro-ecological systems, where similar combinations of market opportunities and resource endowment lead to similar production modes. We distinguish:

(a) grazing systems, where animals get 90 percent of more of their feed from pasture;

(b) mixed farming systems, where animals get at least 10 percent of their feed from crops and crop residues produced on the own farm.

(c) industrial systems, where animals get less than 10 percent of the feed from the own farm.

Globally, mixed farming systems now produce more than half the meat, and 90 percent of the milk, but past trends show a strong growth of the industrial system. However, with globally little additional grazing land available, and also the opportunities for crop land expansion limited the main effect of increasing demand for livestock products would be an increased pressure inside the grazing and mixed farming system.

Grassland-based Systems (LG)

Central and South America and the developed countries dominant in terms of meat production, together accounting for more than three-quarters of the world's production.

Temperate Zones and Tropical Highlands (LGT)

The grassland based system in temperate and tropical highlands is a grazing system constrained by low temperatures. In the temperate zones, there are one or two months of mean temperatures (corrected to sea level) below 5° C, whereas in the tropical highlands daily mean temperatures during the growing period are in the range of 5 - 20° C.

In tropical highlands comprise parts of the highlands of South America and eastern Africa. The cases in temperate zones include southern Australia, New Zealand, and parts of the United States, China and Mongolia. Mongolia's steppe system, New Zealand's dairy and sheep enterprises, dairy systems close to Bogota, Colombia, South American camelid and sheep grazing systems in the Altiplano of Peru and Bolivia. Extensive grazing systems are also found in parts of northwest Pakistan involving sheep for mutton and wool and transhumant sheep on degrading high altitude pasture in Nepal.

Resources and Production

Temperate breeds perform well in tropical highlands situations, except at very high altitudes, only encountered in the Andes of South America. Local breeds play an important role where subsistence objectives are still important, cash income is limited and few purchased inputs are used. At the other end of the intensity scale, New Zealand systems use highly selected animals, artificial insemination (AI) with fresh semen, and a range of advanced technologies to maximise animal output from the pasture produced.

By definition, range is the primary feed resource of this system. Quality varies widely. Oceania's systems involve top-dressing with fertilisers, introduction of legumes, appropriate fencing to achieve highly productive legume-grass pastures. Relatively even rainfall distribution and seasonal mating allow for comparatively high productivity levels using a minimum of hay or silage. In the less intensive grazing systems of Africa and Asia, seasonal fluctuations in feed supply are mainly buffered by the loss of weight of the animals. This, however, limits their productivity.

The developments in New Zealand document the potential of appropriate intensification, which allows for a highly competitive dairy and sheep sector, producing and exporting from a very remote part of the world in spite of heavy subsidies being applied by some competitors. The country has developed labour-extensive, not very capital-intensive technologies to enhance the productivity of the basic resource, a productive rangeland.

In the tropical highlands, the LGT system is affected by seasonality of fodder supply which, in turn, largely depends on rainfall patterns. Extensive systems adapt by accepting weight losses and reductions in milk output. Where milk markets generate the appropriate incentives, dairy cows are either fed cut-and-carry forages or, as is the case in the dairy system of the highlands of Bogota, Colombia, pastures are irrigated.

Product use varies widely, ranging from very commercially export-oriented New Zealand farmers, to South American farmers mainly producing for the domestic market, to Asian and African small-holders catering for local markets and their own subsistence.

Issues and Perspectives

The regions in which the LGT system predominates have a combined human population of 190 million which represents only 3.5 percent of the world total, and almost half of this population lives in Asia. In OECD member countries far fewer people (14 million or 1.7 percent) use the LGT system, but they control more land and cattle per inhabitant than in the other regions.

The major interaction is through the market, where the same animal products are supplied by mixed and landless systems. Market forces and environmental concerns are putting a ceiling on the potential for intensification of this system. Thus, globally, their market share is declining vis-à-vis other production systems.

The major environmental concern in the temperate and tropical highland grazing system is the degradation of rangelands through inappropriate range management practices. These rangelands are frequently part of watersheds, in which range degradation causes problems of flooding, siltation of rivers, etc. Nevertheless, these issues are normally less serious than in mixed farming systems where crop production is practised. Rangeland management frequently involves controlling wildlife, which either compete for forage, transmit diseases or, in the case of predators, cause mortality of animals.

Since the LGT system is found mostly in marginal locations, its production potential in global terms is relatively low. In developing countries it tends to form a subsistence basis for certain groups of the population. Here, its future role is seen more in providing employment for these groups than in making a major contribution to output and economic development. In developed countries, frequently with production surpluses, the production from these systems is declining in relation to other values and uses assigned to these land resources, such as the recreational value, value as a wildlife and biodiversity reserve, and the contribution to water conservation. Therefore, the extensification of production linked with increasing farm sizes may lead to production systems which are both economically viable and environmentally acceptable to societies at large.

Humid and Sub-humid Tropics and Sub-tropics (LGH)

The LGH system is defined as a grazing system found in regions with more than 180 days of growing period. It tends to be concentrated more in the sub-humid zone, particularly in regions where access to markets or, for agronomic reasons, crop production is limited. By definition only very limited cropping is considered for subsistence. The LGH system is found mostly in the tropical and subtropical lowlands of South America: the llanos of Colombia and Venezuela as well as the cerrados of Brazil. Vera and Seré's description of extensive ranching in the Llanos orientales of Colombia at low management and nutrition levels is a typical example. Also, dual purpose (meat and milk) extensive ranching in the Llanos nordorientales of Venezuela based on Criollo, Zebu and European breeds is another typical case.

On the high rainfall side in the humid tropics, the system of Amazonian ranching as well as ranching systems in West and Central Africa developed and belong to this system. At 3540 mm average annual rainfall, beef and milk operations in Amazonia, Colombia and in Brazil are cases in point, producing at low input/low output conditions. Also dual purpose milk-beef systems in the Mexican lowlands and estancias in Argentina are typical cases of this system.

In Asia the development of perennial crops in the high rainfall humid tropics (tea, rubber, oilpalm, etc) and the annual cropping in the sub-humid zones have limited the expansion of pure livestock systems.

In the African setting, many of the potentially suitable land resources are not used as a result of trypanosomiasis constraining livestock production. An example of agropastoralism in the sub-humid zone is described by Otchere for the Kaduna plains in Nigeria. This system is characterised by low performance of cattle due to disease pressure.

Outside Latin America, this system is important only in Australia because of its ample land resources in relation to its population. Worldwide, the LGH system comprises about 190 million head of cattle, an important share of them of Zebu breeds. In the sub-humid and humid regions, cattle are clearly the dominant species, and in very high rainfall areas, such as the Amazon delta and some parts of Queensland, Australia, buffaloes are ranched. African hairsheep and dwarf goats are usually only kept for local consumption. In the sub-tropics, wool-sheep are an important component of the system, for example, in Argentina, Uruguay, South Africa, and Australia.

Herd structure normally reflects the fact that these systems tend to produce mainly beef. Either they sell store cattle for finishing close to market places or they produce finished steers of three to four years of age. Milk is more important in the subtropical and drier parts of the tropics, particularly where farms are smaller and access to markets is provided.

Criollo cattle constituted the main animal resource in tropical Latin America. Over the last 50 years, Zebu cattle have replaced the Criollo cattle in the tropical areas of Latin America. In Australia, British breeds, poorly adapted to the tropical environment have also been increasingly replaced by Bos indicus and its crosses. In Africa, trypanotolerant Bos taurus mainly the N'Dama breed, are important in humid ecosystems, and a range of Bos taurus and Bos indicus breeds are found in sub-humid regions (which tend to be in mixed systems).

As opposed to the LGA system, forage quantity and quality in LGH depend more on soils than on rainfall. African rangelands tend to be of better quality than the Latin American ones. Seasonal fluctuations occur particularly in the sub-humid zone, which are dealt with by exploiting spatial variability of the land resources. The tropical savannas of Latin America are of such low quality that they have to be burnt in order for cattle to graze the young regrowth.

During the wet season, the higher well-drained regions are grazed, while the lowlands are burnt and grazed during the dry season. In the high rainfall rainforest regions, pastures are almost exclusively sown pastures established after clearing. Highest gains are achieved during the relatively dry period when cattle graze forages of somewhat higher dry matter content and animals face less moist soil conditions. Some mineral supplementation (mainly phosphorus) is used given the low fertility and leaching of the soils.

The LGH system produces approximately 6 million MT of beef and veal and 11 million MT of cow milk. By far, the most important geographic region is Central and South America.

Production technology is based on the use of abundant land, some investments such as fencing to improve labour productivity, but very limited purchased inputs and labour. Where milk is produced, inputs and labour are used more intensively. Productivity levels tend to be low (e.g. weight gains of 0.3 kg/head/day in steers, milk yields of 2 kg/cow/day in addition to what the calf suckles). In developed countries, operating with higher product prices, and frequently lower prices for technological inputs, the systems are more intensive and productive in terms of output per animal or per hectare of land.

The LGH system is predominantly market oriented. When distance to the urban markets is large and/or when soils are poor, calves and lean steers are fattened in more convenient systems, leading to a certain degree of specialisation of the systems.

Issues and Perspectives

Globally, 6 percent of the world's population live in areas where the LGH production system predominates. Its importance in terms of sustaining livelihoods of rural populations is expected to decline as interaction with crop cultivation turns it into a mixed system. In rainforest regions, efforts are being made to incorporate perennial tree crops, frequently as silvo-pastoral systems. In the savannas, this system is being converted into a mixed farming system by including annual crops, such as maise, soybeans, sorghum.

The impact of ranching on deforestation of rainforest areas, particularly in Central America, Mexico, and Brazil is one of the more notorious negative impacts attributed to livestock. Particularly in the Brazilian case, this process was fostered by a set of policies, which have been changed in the meantime.

Burning of savanna pastures is another important environmental impact of range utilisation. This is said to release more CO_2 than the burning of tropical forests. Loss of wildlife genetic resources is an issue frequently associated with ranching. In the case of the rainforest regions of Latin

America, where ranching is an important form of land use, this is probably not a major issue, given the large proportion of rainforest still untouched. In the Latin American savannas, human population density is very low and large national parks areas are maintained, again limiting the impact of livestock on wildlife biodiversity.

Horizontal expansion of the LGH system is limited in all agroecological settings. In the rainforest, environmental concerns, technical problems and policies generally discourage further clearing for pasture establishment. In the sub-humid zones, the presence of the LGH system is largely determined by low population pressures, and the existence of lands not attractive for crop production, either because of edaphic restrictions or distance to markets. The transformation of this system into a mixed one is induced by horizontal expansion of crop production driven by population growth and by agricultural research developing crops adapted to the frequently infertile acid soils.

Feed quality is the major constraint in LGH rangelands limiting output per animal. Pasture improvements only play a limited role for solving this bottleneck. The economics of improving pastures are not attractive enough under the prevailing conditions. Both improvements in road infrastructure and new technologies making the joint establishment of pastures with commercially worthwhile nurse crops feasible, are making ley farming systems, involving rotations of crops and pastures, a potentially attractive path into mixed farming systems.

Arid and Semi-arid Tropics and Subtropics (LGA)

The LGA system is defined as a land based system in tropical and subtropical regions with growing periods of less than 180 days, where grazing ruminants are the dominant form of land use, i.e. this enterprise generates more than 90 percent of the total value of production and 10 percent or less of the dry matter eaten by animals is provided by crop production (stubbles, crop by-products or annual forage crops).

This system is found under two contrasting socioeconomic environments: on the one hand, in sub-Saharan Africa and the Near East and North Africa regions, where it constitutes a traditional way of subsistence for important populations, and on the other hand, it is found in Australia, parts of western United States and southern Africa, where private enterprises utilise public or privately owned range resources in the form of ranching. The system is of very limited importance in Central and South America, Asia and Eastern Europe and CIS countries.

Resources and Production

Africa's pastoralists have developed very resilient grazing systems which manage to maintain relatively high human populations on rangelands of low and highly variable productivity. They use a mixture of species (cattle, sheep, goats, camels) and traditional breeds mainly selected for adaptation to the harsh environment. Small ruminants with their higher reproductive rate play a key role in building up livestock populations after periodic droughts have destocked the system.

Under conditions in developed countries, this system has evolved into a very labour-extensive, large scale operation, usually handling only one animal species of a specific breed, particularly Bos indicus cattle breeds. Range is the overwhelming feed resource used in the LGA system. The variation in rainfall quantity and its seasonal distribution determines a high variability over time and space in terms of available feed resources. At the regional level, similar variability can be observed across the country groupings used in this study.

Where the length of the growing season is above 75 days, some cropping is possible. Small areas that can be irrigated are sometimes planted to alfalfa to produce hay to supplement animals on the range. Managing the production risk caused by the variability of feed availability is the central issue in the LGA system with regard to production technology. Pastoralist systems rely mainly on movement of stock across a diverse landscape. In this context the importance of small patches of wetlands and the interface with cropping systems are being recognised. The stocking rates managed by pastoralists under similar ecological conditions are higher than those used by commercial ranchers. This is related to the diverging utility functions of both types of decision makers.

Pastoralists have developed labour-intensive, purchased input-extensive systems. Ranchers use less labour but more capital and tend to make a less efficient use of the land resources, but achieve higher productivity in terms of the livestock capital involved in the operation. They apply a series of management practices and inputs to manage production risk and to enhance output. Many are also related to movement of animals (e.g. transfer of young animals to be fattened in other regions and production systems, strategic stock sales). Price risks are hedged through livestock options markets, particularly in the United States.

Pastoralists basically seek to attain their subsistence from their livestock and are therefore more interested in continuous flows of food, such as milk, dairy products or blood, than in terminal products such as meat of slaughtered animals. This is also a reason for keeping small ruminants, being smaller units easier to handle and trade. With the growing urban demand for livestock products, pastoralists are also increasingly producing for the market.

In ranching systems production is almost exclusively for the market, normally for calves or lean steers, which are then finished in other systems. In the subtropical LGA system wool production plays an important role. This is largely due to the fact that wool production is less sensitive to variations in feed supply than mutton production. An extreme case is the production of Karakul lamb hides in Namibia, where newborn animals are killed for their hides, in this way eliminating the lactation feed requirements of the ewe.

Issues and Perspectives

In western Asia, northern Africa and sub-Saharan Africa, the LGA system is important for the livelihood of large sectors of the rural population. In developed countries this system is extremely labour-extensive.

In sub-Saharan Africa, agropastoralism is the most important interface between livestock production and other agricultural production. In other regions these systems are interrelated with other livestock production systems that have access to better quality feed and are closer to markets. In low income countries without an export market, incentives to produce quality beef are weak. This in turn limits the attractiveness of livestock production stratification.

The degradation of rangelands in the LGA system is an intensively debated issue. The conventional view that population pressure is linked to, the deterioration of traditional property rights, and that the transformation of rangelands into an open resource is leading to their degradation. Degradation is defined along the lines of long term decline in secondary productivity (animal outputs) reflecting declines in primary productivity, as the botanical composition of the vegetation shifts towards less productive and less palatable species and increased soil erosion. This view is presently being challenged by ecologists, who are developing alternative theories of rangeland management, which explicitly take into account the high variability of primary production in these systems.

The evolving views of the functioning of this system are also leading to rapidly evolving development perspectives. It is now acknowledged that pastoralists are making a relatively efficient use of the rangeland resources, although substantial increases in output are improbable and partial interventions are rarely successful. Public sector efforts to manage the system have generally failed. Policies should help pastoralists to be able to operate flexibly to cope with variability. The public sector's role is seen as less regulatory and more into monitoring the situation, to promote efficient use of the rangelands.

Globally, new roles are emerging for these rangelands, besides that of producing ruminant animal products. In developing countries, the pressure to expand crop production is increasing the population pressure on the remaining rangelands. In developed countries, the utilisation of these rangelands for animal production has often been subsidised through very low prices for grazing permits and public investments in irrigation. The rangelands are increasingly seen as a large CO_2 sink, important wildlife habitats, areas of recreational value, etc. Societies will have to find new ways to make the legitimate interests of those presently utilising these resources compatible with the interests of societies at large.

Mixed Rainfed Systems (MR)

Sub-Saharan Africa; West Asia and North Africa; and Central and South America are relatively unimportant in terms of meat production, whereas developed countries and Asia together contribute about 70 percent of the total meat production from mixed rainfed farming systems.

Temperate Zones and Tropical Highlands (MRT)

This system is defined as a combination of rainfed crop and livestock farming in temperate or tropical highland areas in which crops contribute at least 10 percent of the value of total farm output.

The MRT system is found in two contrasting agro-ecozones of the world:

- it is the dominant system in most of North America, Europe and northeastern Asia, basically covering large strips of land north of the parallel of 30° northern latitude;
- it is found in the tropical highlands of eastern Africa (Ethiopia, Burundi, Rwanda) and the Andean region of Latin America (Ecuador, Mexico).

The main common feature of these two regions is that low temperatures during all or part of the year limit and determine vegetation that is quite distinct from the tropical environments.

Typical cases include smallholder peasant farmers in northern China, family-run farms in central and northern Europe as well as North America in the temperate region and typically smallholder operations in the Ethiopian highlands and the highlands of Central and South America (traditional ley farming system with potatoes, barley and pastures as main elements).

Resources and Production

Climatic conditions in temperate regions require substantial active interventions by farmers to feed their animals during the harsh winter period. This necessity makes it economic to maintain animals selected for relatively high levels of productivity. Multi-purpose cattle (meat, milk and traction) were bred over the last 50 years for higher productivity in specialised traits as rapid economic development required increased labour productivity in the rural sector of developed countries. These breeds (Holstein Friesians being a very good example) have been introduced into many livestock systems with very different resources and requirements. Under frequently harsher conditions (lower quality feeds, seasonal feed stresses, higher disease challenge) smaller, less productive but better adapted breeds have developed.

The wide range of intensity levels that are found in this type of system is largely related to the feed resource utilised, and this is in turn largely determined by economic factors, mainly related to the relative prices of livestock outputs in terms of feeds available. (This point is further elaborated in the description of the LLR system). The need to feed animals during winter in temperate regions requires forage conservation. This, in turn, is only economical if the output per animal is relatively high. In countries, where this system is carried out at high levels of intensity, a range of forage and dual purpose crops are grown, frequently in rotation with cash crops.

On the contrary, in the tropical MRT system livestock tend to be of secondary importance vis-à-vis the crops, with animals providing a range of services to the system. In temperate locations, soils play a key role as the major nutrient pool in the system. Over the centuries, farmers have developed a farming system which efficiently recycles nutrients. In the course of the development process, with increasing urbanisation and growing incomes, production technology evolved leading to higher specialisation, more use of external inputs and more open systems. The growing negative externalities of these systems for the environment are inducing important shifts in the type of technologies being developed.

An example is the use of purchased feeds in the land-scarce MRT system of western Europe. This allows an increase in the number of stock kept per farm, thus leading to a manure disposal problem as increasing amounts are spread on a limited farm area. As a consequence water and air pollution are increasing. In response to these problems, a series of regulations and technologies have been adopted to handle animal wastes.

In most tropical MRT systems production is less intensive and soils can cope with higher doses of manure than in temperate regions. In temperate developed countries, the MRT system produces one or a few livestock products almost exclusively for the market. The relative importance of livestock versus crops in terms of income generation grows with rising per capita incomes in the country concerned. At the same time products embody more and more post-production services. On the contrary, in least developed countries (LDC), the MRT system livestock perform a series of functions in mixed systems: a continuous flow of cash income; means to concentrate nutrients through manure (typically potatoes in the Andean MRT system); fuel; animal traction; a cash reserve for emergencies; and as a buffer to risks in crop production.

Issues and Perspectives

The MRT system supports a relatively small and declining number of farmers in developed temperate countries, but additional people are employed as part of the commodity systems (processing, marketing, transportation, supply of inputs, etc). In developing countries, these systems tend to be run by a large number of smallholders. This fact and the similarity of many technical issues to the ones addressed in temperate environments have made this system an attractive subject for technical cooperation projects. Given the magnitude of this system, and the nature of the policies implemented to steer its development, its interactions via markets with other production systems worldwide are substantial.

Over time, the MRT system has been under strong pressure to specialise in meat or dairy production, frequently reducing production of pigs and poultry, where the links to the land are not as direct as in the case of ruminants. This has given rise to the large and dynamic sector of landless monogastrics systems. A similar but less pronounced trend could be observed in the evolution to pure crop production systems and separate landless ruminant systems.

In developing countries with poor road and marketing infrastructure, high capital costs and volatile markets as well as substantially lower opportunity costs for labour, incentives for specialisation are few. Among the land-based systems, the MRT system is certainly the one where technical change has had the largest impact in terms of changes in intensity of production, land use, input use, genetic makeup of breeds, etc. Thus, impacts in terms of loss of domestic animal biodiversity, use of agrochemicals to sustain feed production and waste disposal are substantial.

The MRT system is the largest in terms of stock numbers of both cattle and small ruminants as well as their meat and milk outputs. This can also be expected to link the system to a substantial share of global methane production from ruminants. A large proportion of the land used by the MRT system was originally forest. Clearing occurred over several centuries, but must have made a significant contribution to global CO_2 production. Effluents from tanneries and to a lesser degree from slaughterhouses are an important post-production negative impact of these systems.

Developed countries, where the MRT system predominates, are, as a whole, net importers of many livestock products, among them hides and skins. In the past, these were imported raw and tanned in developed countries. This process is changing with tanning increasingly occurring in LDCs, thus causing the pollution in the countries of origin. Globally, this MRT system is the most important source of animal products, providing 39 percent of the beef and veal production, 24 percent of the sheep and mutton production and 63 percent of the cow milk produced.

The MRT system has rapidly evolved in the past decades in reaction to unprecedented rates of economic growth in the developed countries. This included substantial pressure on the MRT system to increase labour productivity. This was achieved through mechanisation, specialisation, increased use of inputs, and increased scale of operations. The increase in labour productivity achieved was nevertheless obtained at the price of opening the system by importing feeds, fertilisers, fossil energy to operate machines, etc., and extracting large amounts of nutrients via increased outputs or accumulating them as manure beyond the threshold of the quantities which could be efficiently utilised to maintain soil fertility, thus contaminating groundwater, polluting the air, etc.

The challenge for this system is to find ways and means to return to intensity levels with efficient nutrient cycles and fewer negative externalities, while at the same time generating an adequate income for a socially desirable number of farmers, in order to not only to produce agricultural commodities, but also maintain the other functions developed countries' societies expect from these regions (recreational value, clean air and water supply to cities, biodiversity conservation, etc). This will require important changes in policies, institutions and technologies.

In the OECD member countries, such changes could bring a reduction in output levels as prices decline due to the reduced protection of domestic production and increased international trade. In eastern Europe and CIS, mixed systems will replace inefficient large-scale landless systems as these economies open up to the markets. Both developments should introduce more environmentally sound livestock production in the developed temperate economies. This will largely be achieved through the implementation of appropriate policies and regulations. The fact that in developed temperate countries farmers are relatively few, and tend to manage large commercial operations and that the public sector has the resources to enforce policies, makes this avenue feasible.

Humid and Sub-humid Tropics and Subtropics (MRH)

In the humid and sub-humid regions of the tropics and subtropics, livestock production is based on the mixed farming system. Given the range of socioeconomic conditions, soils and climates, this livestock system is very heterogeneous in its composition. It is found in all tropical regions of the world, mainly in developing countries. Parts of the southern United States are the only significant developed region included in this system.

Typical cases are smallholder rice-buffalo systems in southeast Asia or soybean-maize-pasture operations in the Brazilian cerrados of the large-scale commercial type.

Resources and Production

This system includes regions with especially difficult climatic conditions for livestock (high temperatures and high humidity). Adaptation of highly productive temperate breeds to these challenges has been notably poor. In many parts of Africa, trypanosomiasis constitutes an additional constraint for these systems. Particularly in African and Asian smallholder systems, the local breeds are still widely used. In Latin America, Bos taurus cattle, sheep and goats were introduced some four centuries ago. Bos indicus cattle were introduced a few decades ago and have now replaced the earlier introduced cattle breeds in tropical areas.

Among feed resources, the relative importance of rangelands vis-à-vis crop stubbles and straw depend mainly on the relative availability of land. This is depicted by the ratio of agricultural land per inhabitant. In Central and South

America MRH systems comprise five times more agricultural land per inhabitant than in the MRH systems of Asia. This explains the overriding importance of straw as a feed resource in Asia.

The multiple functions of livestock in this system, particularly under smallholder conditions, make technical change very difficult to introduce into this system. Traditional technologies tend to be very efficient in using the local resources, but by definition are neglecting the opportunities created by trade and markets to purchase inputs enhancing the productivity of the local resources. In many cases livestock are only of secondary importance in relation to the farmers' objectives, a fact also reducing the chances for intensification of livestock production.

In Central and South America, extensive ranching systems are increasingly evolving into mixed systems as urban demand for crop staples and livestock products as well as road infrastructure expands. Pastures have traditionally been established jointly with an annual crop, mainly maize or rice. Lack of sustainability of continuous annual cropping on acid infertile soils, typical for the large savanna ecosystems of South America, has increased incentives for developing nutrient efficient crop pasture rotations. This is also related to the shift in policies reducing price support, input subsidies for crop production. Thus a policy framework promoting competitive production systems should also promote more sustainable mixed systems.

In the African and Asian MRH system, the multiple roles of livestock have prevailed, particularly animal traction and manure. In Central and South America, this system caters to large domestic markets and, particularly in the case of Brazil, it is also linked to export markets. Under smallholder conditions milk tends to be a more important output than meat.

Issues and Perspectives

The MRH system applies to approximately 14 percent of the

global population. This ratio is particularly high in sub-Saharan Africa where 41 percent of the region's population is associated with the system and in Central and South America where it is 35 percent.

The system is replacing grazing systems in Africa and Latin America. In Africa, the process is mainly driven by population growth, and, in Central and South America, by economic development and technological innovations. In many parts of the world, farmers are clearing rainforests to expand this system e.g. in South America along the Andean foothills and in Central America. In Africa, this process is somewhat constrained by the tse-tse/trypanosomiasis complex.

This system is particularly important for large areas of sub-Saharan Africa. The main challenge is finding ways to increase the productivity of the system under serious constraints for both public and private investments. It is generally acknowledged that the biological potentials of mixed systems will be the key to productivity increases, and the expectation is that purchased feed inputs will be replaced by enhanced knowledge about the system, in particular, nutrient cycled within the system. In the more humid parts of Asia, annual crops have been replaced by perennial crops and livestock play a minor role.

In Latin America, low population density, high degrees of urbanisation and relatively high per caput incomes have induced farming systems generally more oriented towards livestock production. In the tropical rain-forest regions, very resource consuming systems were established, in some cases driven by policies and in others by poverty. Many of the policies that promoted wasteful utilisation of these resources have been stopped in the process of structural adjustment.

Arid and Semi-arid Tropics and Subtropics (MRA)

The MRA system is a mixed farming systems in tropical and subtropical regions with a vegetation growth period of less than 180 days. The main restriction of this system is the low primary productivity of the land due to low rainfall. The more severe the constraint, the less important crops become in the system and the more livestock take over as the primary income and subsistence source.

The system is important in the West Asia and North Africa region, in parts of the Sahel (Burkina Faso, Nigeria), in large parts of India, and less important in Central and South America. Typical cases are dryland farming-sheep systems in northern Africa and in the Indian subcontinent, and also the small ruminant-cassava systems in northeastern Brazil.

Resources and Production

The more arid the conditions become the greater the necessity to keep livestock as an asset for farmers. Given the low intensity of the system and the multiple purposes of livestock, the introduction of improved breeds has been quite limited. Thus, loss of domestic animal biodiversity is not likely to be very significant under these conditions.

Globally, 11 percent of the world cattle population and 14 percent of sheep and goats are found in this system. Small ruminants are particularly important in West Asia and North Africa under the MRA system. Grazing land not suited for crop production is the main feed resource of the system supported by strategic use of crop stubbles and straw. Land not used for cropping is frequently community owned. Traditional rules on access to this resource have frequently not withstood the changes occurring in the last decades, particularly population pressure. This leads to common problems of overgrazing and resource degradation.

Given the high risk involved in crop production this system tends to produce crops mainly for subsistence. They are usually produced very extensively, thus minimising the financial risks but also limiting the potential for good harvests. Livestock are produced extensively with minimal use of purchased inputs.

As is the case in other largely smallholder systems, livestock have a range of simultaneous roles in this system, including animal traction, production of manure, use as cash reserve, in addition to the production of meat and milk. Fuel-wood is often scarce as a result of deforestation and range degradation, leading to the ever-increasing role of animals as providers of manure for fuel, in addition to means of transport.

Issues and Perspectives

While this system supports larger populations than any grazing system, only 10 percent of the world population is related to this system. Fifty-one percent of the population involved is in Asia, mainly India, and 24 percent in the West Asia and North Africa region. The major concern related to this system is the degradation of land resources, due to their limited production potential under growing population pressure. In livestock terms, this relates particularly to overgrazing and range degradation. This is connected to increasing stock numbers but also to crop production being expanded into increasingly marginal lands.

Given the extensive livestock rearing practised, livestock in the MRA system produce relatively high amounts of methane per animal kept and more so per kilogram of meat or milk produced. There is a close interaction with the LGA system. With increasing population pressure, the LGA system tends to evolve into mixed systems, mainly the MRA system, due to the greater caloric efficiency of cropping vis-à-vis ruminant production when land becomes scarce.

The outlook for this system is relatively similar to the one for the LGA system. The resource base puts a clear ceiling to agricultural intensification. Low and variable response to inputs makes their use financially risky. Population growth in this setting is leading to over-exploitation of the natural resource base, as traditional property rights cannot cope with the growing demands on the resource base. Alternative development strategies and the reduction of population pressure on the resource base are key elements for the sustainable development of these regions.

In the past, irrigation has been seen as the logical strategy to cope with the central constraint to agricultural production in this region, i.e. low and variable rainfall. Results have been mixed at best. Some reasons for the frequent failures were the high investment, the length of the training required to educate rainfed farmers in efficient irrigation management, the short useful life of many irrigation schemes due to salinisation. Furthermore, best locations for irrigation schemes have already been exploited by now. Thus, a blend of other strategies is required in these regions, which involves promoting the mobility of workers to other regions and sectors, the in situ development of other sectors of the economy such as mining, tourism, fisheries, etc.

Mixed Irrigated Systems

The mixed irrigated system contributes about 23 percent of the total meat production worldwide. Draught is another major livestock output.

Temperate Zones and Tropical Highlands (MIT)

This system belongs to the group of the land-based mixed systems of temperate and tropical highland regions. The peculiar feature is the existence of irrigation, which strongly influences the feed availability for ruminants and the variability of crop production. This changes the production environment substantially and determines the competitiveness of animal production vis-à-vis crop production in a given location.

This system is found particularly in the Mediterranean region (Portugal, Italy, Greece, Albania, Bulgaria) and in the Far East (North and South Korea, Japan, and parts of China). These are agro-ecologies in the transition between subtropical and temperate conditions, where plant growth is limited, both by low temperatures in the cold season and by moisture availability during the vegetation period. Their importance in tropical highlands is negligible.

Typical cases are south European family farms combining one cycle of irrigated crop production with livestock production based on the grazing of drylands, crop stubbles and some irrigated alfalfa. The transition to mixed irrigated arid systems is gradual, with the latter having year-round production on irrigated land, thus reducing the opportunities for grazing crop stubbles. Far-east Asian mixed family farms are mainly based on irrigated rice and dairy cattle.

Resources and Production

Traditional local sheep and cattle breeds have been largely displaced as management practices and product prices allowed for more intensive production and the associated increase in the use of external inputs (energy for water pumping, fertilisers, agrochemicals.) In the Mediterranean region, the main feed resource has traditionally been the silvopastoral system, supplemented by crop y-products. In the land-scarce, intensive East Asian systems, the main resources are cereal straw, intensively managed pastures, forages and imported feeds.

Livestock production technology is basically the same utilised by the MRT-system. High product prices and a high opportunity cost for labour make intensive production systems viable. This implies a heavy effort to actively adjust seasonal feed supply to the rather constant requirements of the herds and flocks. This is achieved through forage conservation (hay, silage) and through the feeding of grains and grain by-products.

In the more extensive situations, such as in the Chinese MIT system, the integration of livestock into the farming system is broader in physical terms. Animal traction is an important input into the crop system. Less productive animal breeds are fed less concentrate feeds and therefore consume more crop by-products. Manure is actively allocated to the more productive irrigated fields thus transferring nutrients from other parts of the farm to the irrigated fields. Weeds are fed to the ruminants.

Meat, milk and wool, the main outputs of this system, are mainly produced for the market. Manure is an issue only where animals are stabled, at least for certain periods of the day or the year. Animal traction has been displaced completely by engine-powered equipment in developed countries and the MIT system in China is gradually following the same path. Pigs, ducks, geese and chicken play a minor role, mainly in LDCs in utilising crop by-products and family labour.

Issues and Perspectives

About 10 percent of the world population live in regions where this system is dominant. A large share of them belong to developed countries with relatively high income levels and where agricultural trade is important.

This system tends to be found in regions with rather high population density. The major issue in environmental terms is the use of water, with agriculture competing with the use for urban supply. Another important issue is the management of the lands that are not irrigated. Particularly in the Mediterranean region, complex silvopastoral systems have been developed combining rainfed tree crops (olive trees, hazel nuts, cork-oaks) with extensive grazing, mainly of small ruminants.

Interactions with other systems are mainly trade related and are expected to increase in the future as agricultural protection is reduced. This competition will be mainly with mixed rainfed temperate systems, which produce largely the same commodities.

This system is clearly associated with very intensive agriculture in temperate regions with a high population density. This is the case of the Far East and the southern European regions. They are producing typical commodities of temperate environments at very high levels of intensity. It is related to the historical land scarcity and to policies heavily protecting domestic agriculture. With the outcome of the General Agreement on Tariffs and Trade (GATT) negotiations, it can be expected that these systems will be less and less viable, having to compete with very efficient rain-fed systems producing the same commodities. The system can be expected to shift to more extensive production, using less water and chemical inputs. This will reduce the negative impacts of the system on the environment.

The expansion of international trade and particularly the incorporation of southern European countries into the EU, has led to an increase in the intensive production systems of off-season vegetables and fruits on the best irrigated land. The integration with livestock has been reduced, with ruminant grazing systems declining in absolute terms and concentrating on the marginal sites.

Humid and Sub-humid Tropics and Subtropics (MIH)

This is a mixed system in tropical and subtropical regions with growing seasons of more than 180 days, in which irrigation of crops is significant. The MIH system is particularly important in Asia. High population densities require intensive crop production and the irrigation of rice makes it possible to obtain more than two crops per year, even under conditions of very seasonal rainfall, substantially reducing yield variability as compared with the yield of upland rice or other rain-fed crops. Animal production has in the past been closely linked to the animal traction issue. In many Asian countries, small-scale mechanisation is replacing it now, releasing feed resources for animal production to the markets. Typical cases are irrigated rice-buffalo systems of the Philippines, Vietnam, etc.

Resources and Production

Buffaloes and cattle have mainly been selected for animal traction in this system, involving both tillage and transportation. As mechanisation expands, these animals selected mainly for the adaptation and animal traction performance may gradually be substituted by highly productive breeds to respond to a growing demand for meat and, to a lesser extent, dairy products. Pigs and poultry (particularly ducks and geese) play an important role in utilising otherwise lost feed resources. Potentially valuable genes of adaptation to high fibre diets, tolerance to diseases, etc may be at risk in this system.

Given the land scarcity, the major feed resources comprise crop by-products, straws, brans, weeds and roadside pastures. High yielding varieties of rice have emphasised grain production, frequently at the expense of their contribution to animal feed production (quality and quantity of straw, use of herbicides to control competition from weeds, etc). Highly productive forages for cut-and-carry systems, capable of growing on non-irrigated land are a potential avenue for intensification. Short term forage crops relay-planted into rice fields are also being tested. Tuber crops such as cassava and sweet potatoes, capable of producing acceptable yields of feeds of high energy concentration per kilogram of dry matter are an important resource for pig and to a lesser extent poultry production.

The high productivity of land in this system is achieved through intensive land use of irrigated areas. Hence, the need for animal traction or mechanisation to rapidly till the land after harvest to achieve a new crop cycle. This clearly limits grasing of stubbles and explains the efforts to harvest straw and treat it for feeding ruminants. Cattle and buffaloes are mainly tethered or fed cut-and-carry forages. Ducks are to some extent fed on insects in rice fields, a system in conflict with the increasing use of insecticides in rice production.

The main contribution of ruminants to this system has been animal traction. This function is gradually being taken over by small-scale machinery. Gradually, ruminants are assuming the role of providers of an additional cash income, a way to convert fibrous crop by-products and slack family labour into marketable livestock products, which are increasingly demanded by urban dwellers.

Pigs and poultry provide meat for both home consumption and for the growing urban markets. MIH systems throughout the world produce 13 million tonnes of pork (18 percent of global production), more than any other land-based tropical system. Manure is recycled on the fields.

Issues and Perspectives

Among the tropical and subtropical systems, the MIH system is the one related to the largest population group, 990 million people, 97 percent of which are in Asia. The environmental issues are related to the hygiene risks involved in keeping animals very close to people in areas of high population density. System-wide environmental issues are the frequently low efficiency of water utilisation, and related erosion problems and the production of methane from paddy fields.

Competition for urban markets for livestock products is the main form of interaction with the landless monogastrics system, both domestically and globally through international trade. This system has developed under high population pressure into a very closed system, capable of sustaining the basic needs of a large population. The challenge is how to maintain its sustainability in a changing setting: economic development is creating alternative employment and raising the opportunity cost of labour, consumers are purchasing increasing quantities of animal products and expecting products of different attributes: less fat, more homogeneous characteristics, more processing, etc. At the same time, expanding international trade is providing opportunities to access low cost feeds. These trends are promoting a certain degree of specialisation while environmental concerns favour the maintenance of the traditional highly integrated system.

Arid and Semi-arid Tropics and Subtropics (MIA)

This is a mixed system of arid and semi-arid regions, in which irrigation makes year round intensive crop production feasible. It is found in the Near East, South Asia, North Africa, western United States and Mexico. Typical cases are luzerne/ maize-based intensive dairy systems in California, Israel and Mexico; small-scale buffalo milk production in Pakistan; and animal traction based cash crop production in Egypt and Afghanistan.

Resources and Production

Cattle and buffaloes for milk and animal traction are the main ruminant resource. Sheep and goats are important where marginal rangelands are available in addition to irrigated land. In the MIA system, pigs are kept only in the Far East; they are virtually inexistent in West Asia and North Africa, largely for cultural reasons (Islamic and Jewish religions only). The main introduced breeds are dairy cattle to supply milk to large urban centres. Under good management conditions, intensive dairy schemes have been quite successful in hot but dry environments. Some of the world's highest lactation yields are achieved in the MIA system in Israel and California. The traditional smallholder MIA system in Asia relies heavily on buffaloes for milk production.

Luzerne is the forage crop favoured for use under irrigated conditions, due to the plant's capacity to colonise and improve desert soils brought into irrigation schemes. Furthermore, luzerne is high yielding and of high quality, a fact which makes it particularly suitable to supplement ruminant rations based on straws of low digestibility. Straw from irrigated crops is an important feed resource. In this system, efforts to treat straw to increase digestibility are quite attractive. Under developed country conditions, ample use of concentrates is made to feed high production dairy cows.

Milk production management in the MIA system is highly diverse, ranging from traditional buffalo management in backyards fed mainly cut-and-carry forages and straw to large-scale dairy farms milking several hundred cows, mainly Holstein Friesians. In this case, herd management is aided by computer programmes determining management interventions such as daily levels of concentrate supplementation, timing of drying, vaccinating, pregnancy checking, etc.

In the traditional MIA system, irrigated crop production is the main source of income with livestock playing a very secondary role. This is generally reflected in rather extensive management of the livestock enterprises. Using irrigable land for forage production tends to be economical only for relatively efficient milk production, if an attractive urban market for fresh milk and dairy products exists. This is the case when imports of dried milk and dairy products are restricted or consumers are willing to pay a premium for products made from fresh milk vis-à-vis those based on reconstituted milk. Elsewhere, MIA systems are cash crop oriented and large ruminants are kept mainly for animal traction. Furthermore, fuelwood tends to be a scarce resource in these systems, a fact frequently leading to the use of manure as fuel.

Issues and Perspectives

The MIA system is predominant in regions that are home to over 750 million people, two thirds of them in Asia and one third in West Asia and North Africa. A large proportion of the total labour input into these systems is allocated to irrigated cash crop production.

Milk production is mainly located in the proximity of urban centres. Particularly in modern, large scale operations, manure disposal tends to be an environmental problem. The main draw back of the MIA system, however, is water use, deficient drainage and salinisation of irrigated land. The existence of certain fodder crops that tolerate relatively high levels of salinity, opens an avenue for livestock production as a strategy to live with the problem of salinisation.

The main interaction with other systems occurs through the international market, particularly for milk and dairy products. The MIA system makes an important contribution to food availability and employment in semi-arid and arid regions. The long term sustainability of these systems is nevertheless challenged by the problem of salinisation of soils. Livestock play only an ancillary role, which may even decline in the development process, as appropriate mechanisation becomes economically viable and as freer international trade and better infrastructure enhance the opportunities for consuming livestock products produced within more suitable environments.

LANDLESS SYSTEMS (LL)

The developed countries dominate the picture of landless intensive production with more than half of total meat production. Asia is already contributing some 20 percent and eastern Europe 15 percent, with the latter recently in sharp decline.

Monogastric Production System (LLM)

This system is defined by the use of monogastric species, mainly chicken and pigs in a production system where feed is introduced from outside the farm, thus separating decisions concerning feed use from those of feed production, and particularly of manure utilisation on fields to produce feed and/or cash crops. Thus, this system is an open system in terms of nutrient flow.

The importation of nutrients normally occurs via markets, also international markets. While the return of nutrients through manure frequently causes problems given the high water content and thus high cost of transporting those nutrients to land-based systems capable of using them. On the other hand, mineral fertilisers are frequently a cheaper source of nutrients, thus reducing demand from other production systems for this resource, therefore turning it into "waste". Thus, the disposal of manure creates a major environmental impact of this system, particularly when production takes place close to highly populated urban centres. It also adds dimensions of pollution by odours and human health risks.

Landless monogastrics systems are found predominantly in OECD member countries with 52 percent of the total landless pork production and 58 percent of the landless poultry production globally. In the case of pig production, Asia is second, with 31 percent of the world total. For poultry, Central and South America follow with 15 percent. To a large extent, this geographical distribution is determined by markets and consumption patterns, in addition to urbanisation levels.

In developed countries with abundant road and cooling infrastructure, large-scale landless operations are located close to ports in net grain importing countries such as pig operations in the Netherlands or northern Germany. In grain exporting countries, such as the United States, landless systems tend to be located in grain producing areas, such as the states of Iowa, Illinois, etc. In countries with less developed infrastructures, such as roads and chilling, these operations are close to major urban centres, reflecting the feasibility of transporting grains vis-à-vis animal products.

Resources and Production

The landless monogastric system is almost exclusively based on hybrid and high producing, exotic breeds. This genetic material is widely traded internationally. The expansion of this system is clearly linked with the extinction of traditional breeds.

The system is frequently stratified, implying that different enterprises specialise in the production of parent material, the production of young animals or the fattening process. The short production cycle of these species implies a high turnover and therefore a capacity to rapidly adjust to changes in demand for the products and to prices of inputs. It also implies that stock numbers are a poor indicator for the importance of the sector.

The system is characterised by an ample use of feeds of high energy concentration (mainly cereals, oilseeds and their by-products). This feature is central to understanding the rapid growth of the system worldwide. The high energy concentration allows transport of feeds over longer distances. This provides for the expansion of production pulled by market incentives, based on imported feeds. Production of feeds is separated from their utilisation. Transportation of concentrate feeds can be achieved at substantially lower costs than that of perishable animal products, even though the quantities are larger. Furthermore, consumers tend to pay a premium for fresh animal products vis-à-vis frozen/preserved products. Seasonality of feed production is easily overcome through grain storage and/or deferred purchasing on the market.

The system is very knowledge- and capital-intensive, easily transferred across agroecological conditions, given the scarce links to the land base. Production efficiency is high in terms of output per unit of feed or per man-hour, less so when measured in terms of energy units. Concentrate conversion rates range between 2.5 - 4 kg/kg of pork, 2.0 to 2.5 kg feed DM/ kg of poultry meat, and even lower for eggs. Capital intensity is high in all cases but wide variations are found. Very sophisticated automated systems are used in developed countries, responding to high labour costs. Variability of production within individual enterprises over time is low as long as management systems in place control exogenous factors correctly, i.e. disinfection, isolation from animals external to the system, effective quality control of feed inputs, etc.

Capacity of traditional breeds to cope with these challenges has been replaced by ability to perform at higher levels of efficiency in terms of desired outputs, as long as these external challenges are controlled by management. Management and infrastructure requirements generate large economies of scale in these systems. This implies large herd/ flock sizes, large volumes of wastes and high animal health risks.

Since products of this system are almost exclusively geared to urban markets, they have to comply to standardisation and other specific quality criteria to be efficiently transported, processed and marketed. Many of these criteria are determined by the processing industries, rather than by the final consumers per se.

Issues and Perspectives

Given the tradeable nature of the inputs and the animal products involved in these systems, this system cannot be related to specific populations. Consumers of the system's outputs are mainly urban populations, frequently close to where the production base is located but also in other urban settings due to the active trade.

The large-scale nature of the system and the heavy investments lead to very high labour productivity but very low employment. Thus, this system produces outputs for a large number of urban consumers but generate employment for few people. This employment tends to be relatively stable over time due to the low seasonality of production. While the employment effect at the production level is low, it must be acknowledged that the forward linkages in processing, wholesaling and retailing, as well as the backward linkages in inputs and services required, generate additional employment.

The most important interactions with the environment are generation of large volumes of wastes and air pollution, as well as the increased demand for cereals, with the impact of the latter on the land resource base. In addition, the genetic erosion related to traditional breeds of chicken and pigs is of concern. Finally, given the character of substitutes of ruminant meats, it can be argued that the rapid development of "modern" landless monogastric systems has reduced the market incentives to expand ruminant production, thus reducing pressures for deforestation and degradation of rangelands.

The system is typically competing with traditional land-based production for market shares in the urban markets. It must be kept in mind that poultry and pork are close substitutes for beef and mutton, thus also interacting with the ruminant systems. In a broader sense, the demand for cereals created by these systems is also competing for land resources with land-based ruminant systems.

Given the strong demand for these commodities, production can be expected to continue growing rapidly, particularly in LDCs. Landless poultry and pig production systems account for the majority of the output in developed countries and are rapidly increasing their share in LDCs given their high supply elasticity in the short run. The landless monogastrics system is an open system where important market failures imply a need for regulations. The negative impacts related to waste management are generally clearly located and regulations as well as technological innovations are mitigating the negative effects, particularly in developed countries.

An important trend is the move to select more appropriate sites for production, away from urban centres to where enough land is available to make manure disposal through farming feasible. The environmental impacts of these systems related to their high derived demand of cereals are of a global nature, given the links of these systems to the international grain markets.

Ruminant Production System (LLR)

This system is defined by the use of ruminant species, principally cattle and marginally sheep, in production systems where feed is mainly introduced from outside the farm system, thus separating decisions of feed use from those of feed production and particularly of manure utilisation on fields to produce feed and/or cash crops. Thus, this system is very open in terms of nutrient flow. It shares this feature with the landless monogastrics system. The main difference is that ruminants need more fibrous rations and that the feed conversion of concentrates to liveweight gains is substantially lower. These systems are only competitive under market conditions where consumers can afford to pay a substantial premium for quality beef over chicken or pork.

Milk production has not been included in the quantitative analysis because the border between landless and land-based production is particularly blurred as roughage is essentially required to produce milk from healthy cows. In many cases, transport of roughage over a certain distance is economic. Thus the system description includes considerations on milk production but no quantitative estimates are provided.

Landless ruminant production systems are highly concentrated in a few regions of the world. In the case of cattle, they are almost exclusively found in eastern Europe and the CIS and in a few OECD member countries. Landless sheep production systems are only found in western Asia and Northern Africa.

Typical cases are large-scale feedlots in the United States and in eastern Europe and the CIS. Intensive dairy operations in the same regions are more land-based, due to the need to feed palatable fodder, which cannot be transported economically over long distances. Small-scale peri-urban dairy production, frequent in many LDCs, particularly in Asia, was not included in this system due to its very distinct nature, where manure is frequently recycled to home gardens or used as fuel and feeds which are mainly roughage produced close-by. These types of production are considered under mixed smallholder systems. Examples are sheep fattening in Syria, feedlots in Texas and large-scale dairy operations in Eastern Europe.

Resources and Production

The landless ruminant system is based almost exclusively on high producing, specialised breeds and their crosses which, nevertheless, have not been bred specifically for performance under "landless" conditions. Furthermore, the limited proportion of total animals in these systems indicate that displacement of traditional breeds cannot be attributed specifically to this system.

With regard to milk production, the Holstein Friesian breed is clearly the most important one as well as for beef production, English breeds predominate in the United States, while large European dual-purpose breeds provide animals for fattening. This clearly reflects the overall endowment with land and particularly range. The abundance of rangeland in the United States has led to the specialised production of calves from beef breeds for feedlot operations, while under European conditions these animals are a joint product with milk, mainly from mixed systems.

Apart from the high energy concentration feeds such as grains, this system requires fibrous feeds to maintain the rumen functions. This is frequently achieved through the use of silage, hay or fresh chopped forages. This requirement increases the complexity of these systems. To a large extent, ruminants are used like monogastric animals and their capability of efficiently utilising fibrous feeds, not suitable for direct use by humans, is neglected. This is particularly true for the brief fattening process in North American feedlots, which improves carcass quality of young animals raised mainly on range of low opportunity cost.

This system is producing 12 percent of the global beef production. Production is highly concentrated in developed countries, mainly eastern Europe, the CIS and OECD member countries. The production system is highly capital-intensive, leading to substantial economies of scale. It is also feed-intensive and labour-extensive. Key efficiency parameters are daily weight gains and feed conversion, basically reflecting the efficiency in the use of capital invested in infrastructure or in the form of lean animals and feeds. Weight gains are usually in the range of 1 to 1.5 kg/day and feed conversion rates are about 8 to 10 kg of grains per kg of weight gain.

In market oriented systems, such as the North American feedlot operations, economic performance is largely related to the evolution of prices of lean versus fat animals. Profitability is highest when the price differential for fat animals is large, as this effect is reflected in the price obtained for the total liveweight sold and not only for the additional weight gained in the feedlot. To avoid the downside risk involved in these price fluctuations, feedlot operators often hedge the risk through the option market.

In this system products are almost exclusively geared to urban markets. In the case of the high quality beef produced, there is very limited processing involved. The situation of milk is more similar to the one of poultry and pork, a large and growing proportion being processed into dairy products.

Issues and Perspectives

Direct employment effects of this system are limited. Some additional employment is generated in specialised services and inputs required (particularly feed production, transportation, processing, supply of feeder cattle) as well as in the processing and marketing of the products.

These systems are competing for market shares and resources with all other livestock production systems through the cross-price elasticities for different meats and animal protein sources. Given the dependency on cooling and road infrastructure of trade in fresh animal products, this competition is stronger in developed countries than in countries with poor infrastructures. For the same reason, competition is stronger in cities close to ports than in the hinterland.

This system is closely linked to land-based systems that normally provide the young stock for landless systems. This constitutes an important difference to landless monogastrics systems, in which replacement stock is produced within the same system.

These interactions are basically the same as those of landless monogastric systems. The most important ones are related to the production of animal wastes, leading to water and air pollution, acid rain and human health hazards. Furthermore, these systems induce extensive use of cereals with related environmental concerns (degradation of soils, nutrient transfers, use of agrochemicals, etc).

The landless ruminant system is producing only a small fraction of the ruminant meat output in developed countries and is of negligible importance in LDCs. It is critically dependent on high prices paid for quality beef and milk, and on ample supplies of low cost grains. The landless ruminant system can be expected to continue growing slowly in North America, driven by population growth, but with per caput consumption of beef stagnating. Its importance can be expected to decline in the European Community as production becomes more extensive in response to policies of reducing support to agriculture and promoting environmentally friendlier production systems.

The situation in eastern Europe and the former USSR is different. There, this system developed under central planning as an industrial process to produce these goods. With the shift to market economies, its importance is declining and ruminant production in that part of the world is shifting to the land base and to a smaller scale.

A growing market for grain-fed beef exists in Japan and the newly industrialised countries of Asia. The growth rate of this market will depend mainly on the evolution of the international price of cereals and the increased per caput incomes. This market will in part be supplied domestically and through exports from the United States, Canada, Australia and possibly South America. Globally, the system will continue to be of limited importance and mainly concentrated in the United States and in a few high income, arid countries of the Near East.

Future Trends

Only 9.3 percent of total meat is produced in grassland-based systems, compared to 36.8 percent in landless systems and 5.3 percent for mixed farming systems. One system alone, LLM accounts for more than half (52.3 percent) of global monogastric meat production. Similarly, the MRT system alone produces more than half (55.5 percent) of total milk production from all considered species and 62.6 percent of cow milk production. Egg production is even more concentrated: more than two-thirds of total production is generated by one system, LLM.

The landless intensive systems, particularly the one producing monogastric meat, are the fastest growing meat producing systems. Their growth rate of 4.3 percent p.a. compares with 2.2 for the mixed systems and 0.7 for the grassland based systems. Compared to cropping systems, the system-to-system evolution of livestock systems alone is much less complex. Originally, all livestock production was basically grassland-based. Where climatic, soil and disease conditions permitted, grassland-based systems developed into mixed farming systems which covered a wide range of intensities and production modes as described earlier. The process was basically driven by population density as were the various forms of interaction between the crop and livestock sub-systems. Wherever urbanisation and income exceeded certain levels, landless systems developed in the vicinity of urban centres, capitalising on the efficiency and supply elasticity of these systems.

Similar to crops, two primary sources of production growth can be distinguished: expansion in livestock numbers through an enlargement of the feed resource base, i.e. through an increased intensity of range and pasture utilisation and higher use of feed concentrates and agricultural by-products; and higher output of meat, milk or eggs per animal through improved management, feeding, breeds and animal husbandry technologies. For pigs, poultry and to a lesser extent dairy cattle, much of the increased output comes from landless intensive or mixed farming systems and the use of concentrate. Successes in grain production, on the one hand, and the limited potential for cost effectively increasing quantity and quality of feed from extensive range on the other lead to shift of growth from range to grain-based, production systems, particularly in Asia where growth rates are highest.

The slow evolution from extensive to intensive production has led to increases in environmental degradation, particularly for grasing systems and mixed farming with low degrees of integration. There are considerable institutional and economic problems to be overcome in bringing livestock numbers into balance with forage and feed availabilities. These problems will be difficult to overcome in the short to medium term and are likely to grow in scope and gravity.

Livestock production takes place under very diverse conditions in the different developing countries. However, the direction of change, even if gradual, is towards more intensive production with less dependence on open range feeding which imposes excessive burdens on the environment, and with improved and balanced feeding practices and improved breeds. The improved practices enables more of the feed to go production rather to inefficient maintenance. This has led lead to progresses in feed conversion efficiency.

Intensification must outweigh expansion if livestock production is to respond to effective demand for livestock commodities. Livestock production systems differ in ability to respond to extra demand for livestock products which is primarily due to the biological characteristics of the production process. As an example, poultry meat production has been demonstrated to be quickest in responding to increasing demand.

This is due to the industrial type of production in which this commodity is produced, increasingly also in the developing countries. Fast reproduction cycles allow to react to changing demand within months. A wide range of commodities can be diverted from other uses (food, industrial uses) to feed, if prices permit - this allows for a fast adaptation of the feed resource base for poultry as opposed to ruminants. The demand for land is low which allows the establishment of production units close to consumer centres. High feed conversion efficiencies make poultry production a highly profitable enterprise.

There are other production systems which are also, to a large extent, driven by consumer demand (eggs, then pork, and to a lesser extent dairy) and the size and type of production units for these commodities are largely driven by market forces. The rate of expansion and intensification of these production systems usually does not allow a gradual transformation from traditional production into these modern types. The growth process here tends to be not evolutionary but discontinued.

For red meat, the pace of growth is such that in most cases the evolution of traditional production systems into more intensified modes of production is through adaptation. The type of commodities (beef, mutton and goat meat, and, to a lesser extent, also dairy) lends itself to a more gradual transformation because of the long reproduction cycle, low feed conversion efficiency, and a lower degree of specialisation. This is mainly due to the physical characteristics of the production process.

Traditional, integrated mixed farming systems are also unlikely to be highly responsive to increasing demand simply because of the other functions that livestock have to fulfil within the farm-household system. The production reserves in the form of productivity increases and production enlargement that can be realised through a gradual transformation of traditional farming systems are usually insufficient to respond effectively to growing demand. As a result, almost all developing countries experienced the emergence of modern production systems, similar to those in the developed countries. As traditional systems prove increasingly unable to meet the rising demand, an increasing share of total supply will come from highly intensive systems.

Among the regions, growth rates have been highest in Asia with a 7.8 percent increase annually in beef and veal production, 6.3 percent in sheep and goat meat, 7.0 percent in cow milk production, 7.0 percent in pork, 9.6 in poultry meat and 9.6 percent in poultry eggs. Comparing these growth rates with stock increases where possible (for ruminants) it shows that there is very little horizontal expansion (1.2 percent for cattle stock, 1.8 percent for sheep and goats, 2.3 percent for cattle dairy stocks). This means that annual productivity increases of between 4 and 6 percent have been obtained in the decade. This is unprecedented and can only be compared to the Green Revolution in crop production in the same region during the 1960s and early 1970s.

For ruminants, the growth has taken place almost exclusively in mixed farming systems and landless ruminant systems have not developed to any significant scale. On the contrary, landless monogastric systems are now found in every country of the region and are quickly developing in the vicinity of urban centres. This, in analogy to the growth path precedented by OECD member countries, is driven by increasing per caput incomes, urbanisation and changes in consumption patterns. Unless full sets of regulatory measures are established and enforced peri-urban landless systems will continue to grow quickly, responding to the surging demand of low-priced livestock products.

References

De Boer, J., "Technological and socioeconomic changes: including urbanization, as they impact upon animal production in Asia", *Animal Production and Rural Development: Proceedings of the Sixth AAAP Animal Science Congress*, Vol I. Bangkok, 1992.

Humphrey, J., "The classification of world livestock systems", *A study prepared for the Animal Production and Health Division*, FAO, AGA/MISC/80/3, 1980.

Jahnke, H.E., "Livestock production systems and livestock development in tropical Africa", Kiel, Germany, 1982.

Ruthenberg, H., *Farming systems in the tropics*, Clarendon Press, Oxford, 1980.

Wilson, T., "Integrating livestock and crops for the sustainable use and development of tropical agricultural systems", AGSP-FAO, 1994.

Wint, W. and D. Bourn, *Livestock distribution and the environment in sub-Saharan Africa*, ERGO, Oxford, 1994.

3

Livestock Feeding Strategies

Livestock feeding is a high-risk business. During some years, an operation may not recover out-of-pocket costs. Entry into the livestock feeding business has few restrictions. The cost of feeding per animal drops as the number of animals in the operation increases. Because of the high risks and the economies of scale that favour larger operations, beef-feeding enterprises are not as well adapted to small-scale and part-time farms as are beef cow-calf operations. In addition, less land is required for a livestock feeding operation than for a cow-calf enterprise.

Many livestock feeders purchase lightweight feeder calves, graze them during the spring and summer, and then finish them in the feedlot starting in late summer or fall. Backgrounding is a special type of programme that usually combines pasture systems and lightweight livestock. These livestock require extremely good nutrition, management, and health programmes, but backgrounding can be profitable. Well-managed, high-quality pastures can be used effectively with these lightweight livestock. More information on grazing and backgrounding can be found in Agricultural Alternatives:

The feeding system for a livestock feeding enterprise should remain flexible. For farmer-feeders, corn silage, and occasionally hay crop silages, can be incorporated into the feeding programme. The extent that forages contribute to a ration is determined by the price of feed grains or food processing by-products with equivalent feed value. Increasing forages in the diet of feedlot livestock will generally increase the cost of weight gain (due to slower weight gain and higher interest costs) when grain prices, specifically for corn, are reasonably low. Specific ration composition is determined by the combination of available feedstuffs that will minimise the cost of weight gain, provide a balanced diet, and reach desired endpoints for the market. This feed combination will vary as grain prices change. Performance enhancers such as growth-stimulating implants can also be used. Research has shown that they provide the greatest return of almost any feedlot practice.

Fodder Trees Shrubs

Ligneous plants, which may be trees, small trees , shrubs or undershrubs, are an important component of the fodder resources for livestock and wildlife. The fodder value of their leaves and fruits is often superior to herbaceous plants, particularly in the case of legumes. In arid and semi arid zones, they provide the largest part of the protein supply during the driest months; for example, it is estimated that, in the Sahel, up to 80% of the protein ration is provided by plants of the *Capparaceae* family during the three driest months of the year.

In the past, trees were commonly used as browse and the cutting of leaves or branches for animal feed is recorded as long ago as Roman times. Sometimes ligneous plants were used for different purposes but their importance as browse plants was only recognised late. For example, *Gliricidia maculata* and *G. sepium* were introduced at the end of the 18th century in Africa as shade trees for coffee, tea and cocoa plantations but it was recognised only a few decades ago that they provide a valuable feed, with 20 to 30% N in the leaves, 14% CF, and digestibility ranging from 50 to 75% according to the type of animal. Their main role now is to serve as a supplement to the natural flora (mixed with 50% grass) for feeding livestock, especially during the driest part of the year.

Erythrina burana was used by Central Ethiopian farmers to shade the coffee plantations. It is very common in this part of the world but its properties became known only recently to scientists. The buranas give a much appreciated fodder for the dry season and the leaves, pods and even bark are palatable to animals. The tree is easy to multiply from seeds or cuttings (large cuttings of 2m long and 10 cm diameter are usually used). It is therefore a very much appreciated multi-purpose tree in agroforestry.

Cultivation in Farming Systems

Trees are used both under natural conditions in range areas and under cultivation in farming systems. Trees on ranges are subject to negative natural selection: animals browse the best first. But they also browse mixtures of browse and herbage, which constitute a more balanced diet with fewer risks of poisoning or intoxication. Animals also play an important role in the dissemination of the seeds through the faeces. The key problems are often the increasing rarity of the valuable species, reduced performance of livestock and poisoning due to increased consumption of "useless" browse.

The concept of usefulness is relative. It varies according to conditions but in the natural environment livestock usually select what is good for them. In artificial agricultural systems, one of the key problems is the optimum density of tree plantations in relation to the system—direct browsing or cut-and-carry—and the timing and periodicity of their use, with the associated crops (if any) and the types of animals, etc.

Nutritive Value

A large percentage of, but not all, fodder trees are legumes. Most of the legumes are rich in crude protein but they are not the only plant family which is rich in protein and a high protein content is not the only factor which makes a good fodder plant; the digestibility, which varies with the type of animal, is of great importance.

Leaves and fruits of ligneous species have a much higher level of digestible protein (DCP) than other fodder sources; in Senegal, it is 180 to 200 g DCP/kg DM in browse compared to 100 to 130 for groundnut leaves and 50 to 70 for leaves and fruits of various herbaceous species. Cattle spend a lot of time browsing: 50% in May and June and more than 30% during the rest of the dry season. The regime is properly balanced when there is a sufficient diversity of ligneous species.

Arid Zone

Acacia tortilis

One of the major difficulties with *Acacia tortilis* (Forssk.) Hayne is that the distribution of the species is very extensive and therefore its provenances are quite different. The nomenclature is still somewhat confused and hybridisation exists in the large species. For most modern authors, several species are distinguished:*A. tortilis, A. raddiana, A. heterophylla, A. spirocarpa*. The most elementary precaution in using seeds or material from any tree of the "tortilis" group will therefore be to enquire about the provenance. In addition, some authors have written that the palatability and the digestibility of *Acacia tortilis* (Forssk.) Hayne subsp. *tortilis* Brenan varied considerably accorded to the provenance. Its tannin content may be high and cause toxicity problems.

Semi-arid Zone

Faidherbia albida

The great value of this tree as browse is due to its production of highly mutritive pods which can easily be kept for the dry season but it may have been considered too much as a miracle tree. At least two sub-species are recognised. The production of pods is quite irregular but a tree may give up to 350 kg in a good year. Gassama has developed a technique of micropropagation from nodes collected on suckers.

Halophytes

In dry areas, whether warm or cold, halophytes are common on salty soils. They have a greater biomass than non-halophytic plants. In Algeria, *Atriplex halimus* produced 3.7t

DM/ha compared to 0.8 to 1t DM/ha with *Artemisia herba alba* or *Arthrophytum scoparium*. In 1988, Iran had already planted more than 43 000 ha of rangelands with *Atriplex canescens, A. lentiformis* and *Haloxylon persicum,* so doubling the fodder productivity of its arid and semi-arid territory.

Climatic effects

In Northern Senegal, Poupon found that climatic variation may have an important effect on ligneous plants. During the 1972 drought, foliation was delayed by 4 to 7 weeks and its intensity was reduced; leaves did not regrow after they were browsed and the majority of flowers did not take shape, resulting in a very low and negligible fruit production.

In the semi-arid zone, there are relatively few species, but a large percentage of them is used as fodder and especially as browse by the animals. Wildlife is relatively more important, at least potentially, then domestic livestock. *Acacia* and related genera are the most important browse species, with most of the Capparaceae in the arid zones themselves, together with other genera such as *Balanites* and *Albizzia*. Le Houérou has established that in *grosso modo* in the Mediterranean basin each millimetre of rainfall was producing 4 kg of aerial biomass or 2 kg of palatable DM or 0.66 feed units (FU) and that in the Sahelian zone the production of one millimetre of rainfall is roughly equivalent to 2.5 kg of aerial biomass or 1 kg of palatable DM or 0,40 FU.

Sub-humid Zone

Leucaena leucocephala

Although this small tree has been used with success in semi-arid zone, it is in the humid and semi-humid zone that it gives the best results. Like the two other species already briefly considered, the problem of lineage is very important. Many varieties have already been isolated and are being used. The big danger with the species is its extremely broad distribution through agroforestry and other development projects, which makes it a particularly sensitive species to diseases and parasites such as psyllids.

The presence of mimosine in several cultivars has created problems with some animals, now better controlled by "vaccination" with rumen fluid from resistant livestock. The odour and taste given to milk of animals fed with *L. leucocephala* may be avoided easily by preventing the animals from eating *Leucaena* for two hours before milking. Inversely, the quality of the meat may be influenced by the nature of the feed: *pré-salé* mutton, produced for example in the Bay of the Mount Saint Michel in France is an example.

The toxicity of mimosine in *Leucaena* is comparable to many other similar phenomenons caused by the consumption of legumes and other browse. *Gliricidia sepium*, for example, is reported in some parts of the world as having an anti-nutritional effect: livestock are fond of it but do not grow when eating it even in large quantities. Perhaps inoculation of rumen fluid from resistant animals could help to cure this. These antinutritive substances seem to appear in some geographical regions or with some specific races or lineages of animals. It is embarrassing when some provenances of *Gliricidia sepium* were recognised as giving as much as 12 to 20 kg DM/tree/year in Maseno (Kenya).

These production figures are surpassed by the production of *Calliandra calothyrsus* which, in Yaoundé (Cameroon), reached 4.69t DM/ha at one year of age with 10,000 trees/ha when *Leucaena leucocephala* was giving 3.78t DM/ha, *Gliricidia sepium* 2.3t DM/ha and *Racosperma auriculiforme* 1.56t DM/ha.

Cutting ligneous vegetation may sometimes increase the total available biomass; this is why the American and Australian ranchers have, for decades, suppressed ligneous vegetation as much as possible from their ranches. but other research has shown the positive action of ligneous vegetation on the total biomass productivity and on its qualitative amelioration.

Humid Zone

Guilelma gasipaes

The peach palm *Guilelma gasipaes*, is a real multi-purpose tree species. In addition to providing the Indians of Amazonia with a fruit which gives oil and starch, it is also important source of browse for the animals. The fruits, given mainly to pigs and poultry, are too small for human consumption, damaged or have some malformation. They are left on the tree until they become overripe, otherwise it has to be cooked as it contains a diastasis which inhibits an other proteolytic ezyme which is necessary for the proper utilisation of protein by animals.

Jessenia battaua

The management of the bataua plam,*Jessenia battaua*, from South America, which exists only for the time being in wild state, seems to be theoretically comparable. But the value of its oil, which costs ten times mote than the oil of the oil plam tree, could modify its potential for management.

Browse Trees

Goats and camels feed essntially on ligneous plants: camels in Northern kenya spend 50% of their grazing time in browsing plants at a height of over 1.6m. There are ligneous plants which provide important quantities of fodder although their main importance lies in other products. In the Salmé area in the medium range elevations of Nepal, forest and wild trees provide the most important part of available fodder. Each tree traditional rules for the period and the frequency the tree is to be lopped for browse. Sometimes the products of a cutting are for sheep, sometimes for goats, sometimes for cattle, In addition, there are trees and shrubs elsewher which are specially protected, or even planted and cultivated, for honeybees, silkworms or other small animals; the economic returns from these small animals are usually greater than those from larger one.

In the Chinampas of Mexico, trees and vegetables are grown on strips of land. Many of the trees are multi-purpose species, but most of them produce fodder. A similar system is in use int he eastern marshy lands of China; here, on the strips of land, livestock (usually goats) are partly bred and fed on the foliage of the trees and shrubs, mainly *Populus* and *Salix* species.

The reson for breeding livestock in such this environment is to provide the soil with manure, which is necessary for the production of vegetables and mushrooms. In the Ban Tré Province of Vitnam, a comparable system has been studied where th leaves trees are used to feed fish.

The oldest agrosilipastoral system recorded in the world, the combination of olive tree, cereal and sheep breeding, is common all around the Mediterrance. From the browse and fodder point of view, the olive tree not only provide browse whtn it is cut (twigs and leaves) but it favours the growth of finer and tender grass in its shade.

In addition to their role in priduction of browse, trees have other production and service role. They produce wood for fuel and building purposes. They give food, fibre, drugs, fruits, pollen, nectar, dyes, gums, waxes, resins, etc. They play a very important role as wind breaks, in providing shade and protection against heat and cold and in reducing erosion.

Crop Residues

This system will have its application in regions of very high population density where crop production is the predominant activity and the major feed resources are the residues and by-products of food crops.

Straws from Cereal Crops

Balance of nutrients

Crop residues and fibrous agro-industrial by-products are characterised by low to moderate digestibility, and low levels of nitrogen, protein and minerals. Classical temperate country wisdom diagnosed energy density as the first constraint and the proposed solutions were supplementation with energy-rich feeds (e.g., cereal grains, root crops or specialist forage crops for feeding directly or after ensiling). Even when it was recognised that these strategies were inappropriate and uneconomic, the classic approach persisted and scientists sought solutions through chemical treatment of the fibrous feeds to improve digestibility.

There is a definite place for chemical treatment of fibrous crop residues but the first step must be to correct the balance of nutrients, since this is the first constraint to productivity of livestock fed these resources.

The system involves:

— Feeding the rumen microbes

— Feeding the animal a source of by-pass protein

Feeding the rumen microbes

There are three approaches to providing the rumen microbes with a better array of nutrients, including fermentable energy:

— Supply critical nutrients (missing in the residue) by means of a multinutritional block.

— Facilitate selection by the animal of the more digestible components of the residue by offering at least twice the amount it can be expected to consume.

— Saponify partially the phenolic linkages with acids or alkalis.

The choice of method will be determined by the relative availability of the residue, alternative end uses (e.g., as fuel or as roofing material), the cost of the urea (or ammonia) and the convenience of the technology. Provision of multi-nutrient blocks has had the widest impact. The selection approach has advantages when there is a need for feed and fuel, since the parts of the residue rejected by the animal are those which are drier and more lignified and therefore of higher fuel value. Saponification with ammonia accomplishes two tasks. It supplies ammonia and also energy (by breaking phenolic linkages in the cell wall) to the microbes.

Multi-nutrient blocks

Fibrous crop residues and agro-industrial by-products are the typical diet of ruminant animals in most tropical countries and often are the only feed resource in extended dry seasons. These feeds are characterised by an imbalanced array of nutrients, of which fermentable nitrogen is usually the first limiting; organic matter digestibility is also usually below 50%. The use of solidified blocks containing urea, minerals (and often rumen by-pass nutrients), pioneered by Professor R. A. Leng of the University of New England, Armidale, Australia, to supplement these fibrous feed resources has been outstandingly successful in a large number of countries.

The major problem with the technology has been the difficulties in some situations of making blocks of sufficiently hard consistency to limit intake and ensure there will be no risk of urea toxicity which might occur if intake of the block is excessively high. Molasses, which is the usual basis of the multinutritional block, is also not always available or it may be too expensive through high transport charges.

Ammoniation

Alkali treatment of fibrous crop residues is well researched and the possibility of using urea as a source of ammonia led to expectations of rapid implementation in many developing countries which, however, have not been realised for several reasons. Too much attention to treatment technique *per se*, and too little to supplementation strategies aimed at employing biologically effective supplements available to the farmers, are some of the explanations. Unsupplemented ammoniated straw supports production levels far below the level potentially made available by the increase in digestibility and intake due to ammoniation. Low levels of production may not earn farmers sufficient income to pay for the ammoniation treatment and, consequently, they lose interest and the technology is not taken up.

Lack of feedback from extension or no communication at all between research and extension can be mentioned as some of the other reasons for lack of impact of new technologies. There has been insufficient research with a farmer's perspective in the area of crop residue utilisation although good extension work must be based on precise knowledge.

It is generally believed that the response to ammoniation has two components: an increase in digestibility due to partial saponification of the lignin-cellulose-hemicellulose linkages, and a greater feed intake arising from the greater supply of ammonia to the rumen micro-organisms. Ammonification supplies the nitrogen needs of rumen microbes as well as increasing digestibility; however, it is an expensive way of supplying the nitrogen, as the level required for effective treatment of the residue is some 50% greater than what is needed by the rumen microbes.

An important economic issue, which has not been adequately evaluated, is the relative effectiveness of ammoniation using urea as compared with supplementing untreated straw with a molasses-urea block. Another issue that requires clarification is the optimum level of urea for effective ammoniation. The usually recommended level is 5% of the dry weight of the residue; yet researchers in Vietnam contend that a lower urea level of 2.5% urea complemented with 0.5% of calcium hydroxide gives almost as good a biological response and is more economical. As the price of urea is tending to rise, as Governments in developing countries reduce agricultural subsidies, it is opportune to examine more closely the possibility of using lower inputs.

Strategic Supplementation

As with untreated crop residues, appropriate supplementation is the key to securing adequate animal response. Mostly protein-rich by-products from oilseed or cereal grain milling have been used as the source of "by-pass" protein. However, there is much potential in the use of leaves from legume trees for this purpose.

The experience in India by the National Dairy Development Board in their rural development programme "Operation Flood" affords an example of the impact of replacing inappropriate "temperate country technologies" (supplementing crop residues with "balanced dairy concentrates") with appropriate technologies (multi-nutrient blocks and "by-pass" protein) based on the concepts of strategic supplementation of rumen microbes and the animal.

Stovers from maize and sorghum

There are few reports of uptake by farmers of the ammoniation technique applied to stovers derived from maize, sorghum and millet, although technically it has been shown to be feasible but more difficult than for straws, due to the bulkiness of the material. The component parts of these residues can differ quite widely in digestibility and it would seem more appropriate perhaps to use the "high-offer" feeding system to facilitate selection of the more digestible components rather than urea treatment.

Residues from plantains and bananas

In some tropical regions, plantains and bananas are the staple of the human diet (e.g., in the Kiliminjaro region of Tanzania). The principles for using these as the basis of the diet of ruminant animals are the same as for cereal stovers. The bulk of the biomass residue is in the pseudo-stem which is of low nitrogen content but highly digestible, while the protein in the leaf lamina appears to be largely bound to tannin-like substances giving it a low digestibility. Therefore the supplements needed are:

- Multi-nutritional blocks to provide urea and minerals;
- A source of by-pass protein (from oilseed meals, rice polishings or as leaves from legume trees).

Residual Fibre

Sugar factory bagasse

The residual fibre after industrial separation of sugar by repeated extraction is described as "factory" bagasse to distinguish it from that produced in simple 2- or 3-roll crushers (*trapiches*) which also contains residual juice (from 20 to 30% of the dry matter). Factory bagasse contains about 45% cellulose, 35% hemicellulose and 10% lignin. The raw material has a very low digestibility (about 30%). However, dramatic improvements can be brought about by high pressure steam (13 kg/cm^2, 200°C) which, through acid hydrolysis (acetic acid is generated in the process), solubilises the hemicellulose component.

Steam hydrolysed bagasse must be supplemented with urea and by-pass nutrients after which it supports quite high levels of production in fattening cattle. This technology has been applied commercially in intensive cattle fattening in Colombia and Brazil. The Colombian programme has been discontinued in favour of processing the bagasse for paper. However, in Brazil where the integration of sugar and alcohol production results in large surpluses of bagasse it is reported that more than 200,000 cattle are being fed with steam-hydrolysed bagasse.

Alkaline hydrolysis of factory bagasse with sodium hydroxide is also an effective method of upgrading the bagasse and has been employed in Cuba on a large scale, although mostly with the short fibre fraction (*bagacillo*) which remains when the larger fibres are separated out as raw material for paper production. The method is relatively costly, as well as polluting because of the large quantities of sodium hydroxide required. It is not a sustainable technology and therefore is not recommended.

Rising oil prices and increasing awareness of environmental issues will almost certainly favour the future use of bagasse as fuel with electricity generation as the major output, in co-generation schemes which will also produce sugar, alcohol and/or molasses from the cane juice.

Sugar cane tops and pressed sugar cane stalk. The fibrous residues derived from farm scale fractionation of sugar cane, and from artisan production of "*panela*" or "*gur*", which have potential for feeding to herbivores are:

— The cane tops

— The pressed stalk

Cane tops

When complemented with rumen and "by-pass"

supplements, cane tops will support liveweight gains of the order of 700 g/d in fattening cattle and 3,000 litres lactation yield in milking animals. It is important to encourage selection by offering quantities of tops which exceed by at least 50% the expected intake. Under these circumstances the animal has been observed to preferentially select the growing point of the cane and to reject in large part the leaf blade.

The justification for this selection can be seen in the data for dry matter degradabilities in rumen nylon bags. The growing point has a much higher rate of degradability than the leaf blade, and is comparable to that of nutritious grass. The African hair sheep appears to be a particularly appropriate target animal for utilising the cane tops. Since late 1989, some 100 breeding ewes and their progeny have been managed in full confinement on this system in Colombia. They have free access to cane tops, multinutritional blocks and a 9:1 mixture of poultry litter and rice polishings. The advantages of sheep in such a system are:

- Low investment in animals and housing (renewable bamboo poles and palm leaves). No concrete is needed as the rejected fibrous components of the feed are used as built-up litter.
- The sheep are highly resistant to common diseases and the housing system reduces risks of endo- and ecto-parasite infections.
- Sheep are selective feeders and adapt more readily to confinement than goats.
- Sheep production is especially suitable for cash-poor smallholder farmers since:
- Investment risk is distributed among many animals rather than in few, as would be the case for cattle.
- A lamb can be consumed by a family in 1 or 2 days, avoiding need for refrigerated storage, or it (or a sheep) can be sold for cash. None of this is possible with cattle.

Pressed sugar cane stalk

The pressed stalk which remains after a single-pass extraction of the cane stalk through a 3-roll mill (or 2 to 3 passes in a mill with only 2 rolls) contains some 20–30% of soluble sugars in the dry matter. The associated cell wall material is, however, of low digestibility as it has the same composition as sugar factory bagasse which is only about 28% digestible.

The pressed stalk is particularly appropriate for manipulating via the high-offer level feeding system. When it is coarsely chopped and offered at 2 to 3 times expected intake, it is relatively easy for cattle and sheep to select the more digestible component (the soft pith, rich in sugar). This feed should be supplemented with multi-nutritional blocks, fodder tree foliages (about 2–3 kg fresh material per 100 kg liveweight) and restricted amounts (500 g/day) of a supplement rich in by-pass protein such as rice polishings. Growth rates of Fl crossbred from weaning to point of calving on this feeding system have been of the order of 500 g/day. Fattening bulls have gained at similar rates when supplemented with *Gliricidia sepium* foliage, a 20% urea block and 1 kg/day of a mixture of poultry litter and rice polishings.

Crops with High Yields of Biomass

The concept of tropical biomass and integrated farming systems as the basis of sustainability. The essential features of a strategy to optimise production and use of natural resources is to ensure that the biomass can be processed in a way that will contribute efficiently to provision of human food, animal feed, chemicals and energy. The animal feed element must be further defined so as to take account of the differing needs of monogastric and herbivorous animals.

The corollary to fractionation is integration, since in order to efficiently utilise the different components of plant biomass (e.g., soluble cell contents, cell wall material with different degrees of lignification), it is essential to tailor end uses accordingly. Thus it makes no sense to feed sugar cane juice or bananas to herbivorous animals when much higher biological efficiencies can be obtained by using them as the basal diet of pigs and poultry. Similarly, it will be more economical and ecologically advantageous to convert sugar cane bagasse, tree branches and highly lignified crop residues into fuel than to process them for feeding to herbivorous livestock. Wherever possible the farming system therefore should offer opportunities for both monogastric and herbivorous livestock production, and for generation of energy.

Thus, although it is feasible to feed a wide range of tropical high-biomass crops, including whole sugar cane, to ruminant livestock, such measures should be viewed as short term solutions. The correct long term approach must be to fractionate the crop according to the most profitable end-product possibilities and to manage and use them always with the aim of optimising the productivity of the whole farming system.

Sugar Cane and its Derivatives

Sugar Cane as Animal Feed

An important advantage of sugar cane as a multi-purpose crop is its beneficial effects on soil fertility. Thus in the acid-sulphate soils region in the Mekong Delta in Vietnam sugar cane is grown in short rotation with rice (8 months cane: 3 months rice) precisely because the rice crop (for food security) would have very low yields if it were not preceded by the sugar cane. In addition, the farmers inter-cropped the cane during the first 3 months with legume beans; the reason was that, in association with cane, there was negligible pest damage on the beans, compared with growing the beans as a single crop.

Maintenance and, even better, the capacity to increase soil fertility will become increasingly important in the choice of crop rotations. Thus, sugar cane has potentially many advantages as a component of cropping systems in the tropics. Management systems should be aimed to exploit these potential assets.

The following criteria can serve as a guide in managing cropping systems based on sugar cane:

— Selection of varieties which optimise yield of biomass, pest resistance and ease of fractionation, and not necessarily percentage of sucrose.
— Intercropping with legume grain crops to provide supplementary protein (e.g., soybean and peanuts), to increase biodiversity and improve soil fertility.
— Alley cropping with different species of multi-purpose trees (e.g., 20– 30m strips of trees alternating with the cane) to provide additional protein-rich feed, fuel and timber, erosion control and biodiversity.
— Fractionation as the preferred processing method to optimise use of the different components.
— Use of the tops and the pressed cane stalk as feed for ruminants.
— The juice for pigs and water fowl.
— The pressed cane stalk for fuel (with or without selection by ruminant animals).
— The leaf trash as an energy source for soil micro-organisms.

Chopped whole and derinded sugar cane

The idea of separating the cane stalk into rind and sugar-rich pith, using sophisticated wood processing technology, was promoted strongly by Canadian developers in the late '60s and early '70s. The aim was to transform the rind into paper and compressed board and use the residual pith as animal feed. Although theoretically superior to chopped whole cane, in practice the advantages of the derinded cane over chopped whole cane were insignificant biologically and were more than outweighed by the high investment and operating costs of the derinding equipment.

Research on chopped whole cane as ruminant feed emphasised the important role of urea, of by-pass protein, starch and oil and of using sugar cane with high content of sugars. Rice polishings—which are relatively rich in well balanced amino acids, starch and oil—have a physical form which facilitates almost complete escape from the rumen. They have proved to be the most effective supplement when sugar cane and its derivatives are fed to ruminants. They are widely available in most tropical countries.

The unique characteristics of sugar cane enabling it to reach its maximum nutritive value in the dry season, has made it an attractive complement to pastures which are low in both quantity and nutritive value in the dry season. Restricted feeding of foliage from leguminous trees (*Leucaena* and *Gliricidia*) has proved to be a cost-effective partial replacement of the rice polishings in these systems.

However, as has proved to be the case with final "C" molasses, so with whole sugar cane, low profitability of beef and milk production has led to new developments in its use which in the long term promise to bring greater cost benefits. In most situations today, fractionation of the cane stalk into juice and residual pressed stalk for use in integrated farming systems, is proving to be a more attractive strategy with which to face the challenge of making more efficient and sustainable use of tropical biomass resources.

Final molasses

Methods for using high levels of "C" or final molasses as the basis of intensive cattle fattening were developed in Cuba in the late'60s. These first approaches used fish meal as the source of by-pass or escape protein. This was the first large scale commercial application of the concept of by-pass protein feeding, hypothesised originally in 1963.

Subsequent work focused on use of cheaper alternatives and led to the idea of supplying both roughage and by-pass protein in the form of crop and tree foliage. Increased costs of molasses and decreasing profit margins for intensive beef fattening in developing countries have, in most situations, made high level use of molasses uneconomical. Emphasis is now on using molasses as a vehicle for urea and minerals in the form of liquid mixtures or as solidified blocks. The latter technology has proved to be particularly attractive and, following its successful introduction in India, it has been transferred successfully to many tropical countries.

Livestock Production and Pasture

Grazing Animal and Sustainable Use of Natural Resource

In Africa, transhumant pastoralism was a life style, that has gradually become less sustainable due to the impact of "development". The upset of the traditional system, and the imbalance caused in the natural wild life, is what has led to the present widespread crosion and desertification. Supervised grazing has succeeded in Asia because it is highly integrated with crop production which is the dominant activity and there is no conflict. It is a threat in Africa, partly because it was the dominant activity and has come under threat by crop farmers; in Africa, grazing is a source of conflict. In Latin America, ranching is also a life style—witness the highly successful (though morally questionable) advertisement for a successful brand of cigarettes. More than any other human activity, in that continent, it has been responsible for immense ecological and social damage.

Pasture can be a sustainable land use system only if it is a by-product of food crop production (the Asian model) or is closely integrated with forestry which should be the dominant activity (agroforestry). Interestingly, when the Asian model was introduced as the alternative (new) livestock production system in the previously eroded "Hado" region, the results were highly successful and have been sustained.

Agroforestry has been practised for many decades in industrial plantations of coconuts, rubber and oil palm, primarily as a means of controlling weed and grass growth. The limited technologies introduced consisted of planting aggressive legumes such as tropical kudzu (*Pueraria phaseoloides*), but here the aim was more to aid the tree crop than the animal.

The "Alley farming" system of agroforestry developed in West Africa was never intended for grazing animals but for "cut and carry" management, or simply as a source of mulch and fertiliser. The planning of leguminous trees in association with pasture for sustainable production of animal feed is a new endeavour (Molina, C. and Molina, E. 1994, Personal communication). In this system trees such as *Gliricidia sepium, Leucaena leucocephala* and *Erythrina fusca* are planted at densities in the range 600 to 1100/ha (*E. fusca*), 10,000 to 20,000 (*G. sepium, L. leucocephala*) and 25–50/ha (*Prosopis juliflora*), in association with grasses such as star grass (*Cynodon nlemfuensis*) and Argentina grass (*Cynodon dactylon*). The trees are lopped at intervals of 90–120 days in the case of *E. fusca* and *G. sepium*, browsed at intervals of 40–60 days for *L. leucocephala* or left for the fruits to fall and be consumed in situ or collected (*P. juliflora*).

Some recent data on soil fertility in agro-forestry systems are interesting and extremely relevant to the issue of sustainable use of natural resources. In the examples, one of a protein bank of *G. sepium* and the other with cattle grazing under *E. fusca*, soil fertility improved over time or in comparison with similar pastures not associated with trees.

Effects of shade

There is a wealth of literature on the composition of grasses, and how this changes under the influence of management including, cutting, grazing, fertilisation, and inter-sowing with herbaceous legumes. There is much less information on the effect of shade on pasture productivity and quality. Work done in industrial tree crop plantations indicates that increasing degree of shade reduces biomass productivity, increases the content of soluble nitrogenous compounds and decreases that of soluble carbohydrates. These changes are likely to have negative effects on both stocking rate and balance of nutrients.

There are no comparable data for associations where leguminous tree crops and grasses are managed for animal feed and where shading is cyclical (i.e., zero immediately after lopping increasing over time to almost 100% shade, depending on the tree population and the harvest interval, which is usually between 90 and 120 days).

Supplementation

If the pasture is young (less than 30–40 days regrowth) then in all probability it will contain adequate amounts of fermentable nitrogen, and minerals and little or no "by-pass" protein. It will also (pasture under shade) be deficient in glucose precursors due to low levels of soluble carbohydrates.

The first priority, in order to increase animal productivity will be the provision of "by-pass" protein. This could be provided by the leaves of the associated trees; otherwise, protein-rich meals from oil or cereal milling should be given. In Mauritius, where dairy cattle are kept in confinement by landless farmers (mostly women), and feed must be harvested from roadsides or sugar cane fields, supplementation with cottonseed meal at only 250g/litre of milk gave the same yield as twice the amount (500 g/litre) of "balanced" concentrates.

The second priority will be to increase the supply of glucose precursors. By-product oilseed meals and brans will normally increase the supply of glucose precursors, either via "by-pass" starch or indirectly through increasing rumen propionate.

A more interesting approach, and in line with the strategy of identifying and using truly tropical feed resources, is the use of the crude oil from the African oil palm. "By-pass" oil (protected with calcium salts) is incorporated directly into milk and body fat, thus saving glucose needed for NADPH synthesis when fat is synthesised from acetate. This feed resource, which is already being used commercially for feeding to pigs, on several farms in Colombia, promises to have a particular role in complementing tree foliages in the diet of milking cows.

Of special interest is an apparent synergistic effect when the oil is mixed with protein-rich leaves from forage trees which normally are not relished by cattle. The fresh leaves of the tree *Erythrina fusca* are eaten by cattle but not with great relish. Wilting them for 24 hours improved both intake and growth response in crossbred heifers. However, mixing the leaves with 6% palm oil (fresh weight basis) and 2% calcium hydroxide, following wilting, brought about a threefold increase in intake.

Strategies and innovations for Animal feed Improvement

Because of the growing demand for animal feed, new technologies and techniques need to be continuously developed and transferred in order to avoid environmental deterioration or increases in the prices of food products. Research and technology generation seeking ways to overcome food insecurity and poverty are therefore essential for agricultural and rural development.

In the development of research strategies and innovations for animal feed improvement and technology transfer methodologies, the following considerations should be made:

— Appropriate technology should be developed to fit the livestock production needs of the rural poor. Accordingly, high priority should be placed on the direct involvement of research beneficiaries, individually and collectively, in the selection of research priorities, the conduct and ongoing evaluation of research programmes and the validation of research results.

— Feed research in developing countries should be based on the specific conditions of each agro-ecological zone. Different ecological zones (such as arid, semi-arid and sub-humid, highlands and humid zones) will require different research-based approaches.

— On the other hand, selecting hardware and technology merely because positive results were obtained under similar ecological conditions is not a guarantee to success. Other factors play a pivotal role in the adoption of new technology, including socio-economic factors (e.g. land ownership, labour requirements and lack of knowledge), cultural factors (e.g. traditional practices and indigenous knowledge), appropriate training support and availability of inputs and technical follow-up.

Appropriate research strategies should therefore seek to identify the inputs, outputs, potentials and constraints of the research component; develop interventions to resolve constraints and exploit potentials; design alternative systems; evaluate interventions and alternative systems; and elaborate alternatives that have proven to be socially and culturally acceptable, technically practical and economically feasible.

Support to research and transfer of technology is critical to the development and sustainability of smallholder livestock production. The Fund has also learned that the development of improved feeding systems is constrained by inadequate technical, economical and institutional capacity in the field of research.

The improvement of national research institutions is essential to the development of sustainable livestock production. Institutional capacity-building is usually achieved through partnerships with local organisations, non-governmental organisations (NGOs), private firms and other entities involved in the technology development and implementation process. The aim is to increase the productivity and effectiveness of the resources devoted to agricultural research.

Building on the complementary relationship between livestock and crops, the Fund has shown, particularly through the introduction of mixed farming systems, that smallholder agricultural systems can be improved in a sustainable way. Integrated crop-livestock farming systems not only play an important role in the provision of by-products that serve as inputs (such as crop residues and manure), but they also reduce soil degradation and are therefore environmentally beneficial. An integrated approach results in increased production, greater income opportunities, and improved diets through the addition of protein, thereby contributing to the sustainability of smallholder livestock production.

In areas where climatic conditions are extreme (e.g. harsh winters or recurrent drought), poor households are often exposed to feeding problems. Because of communal ownership of grazing lands and complex grazing rights in areas with extreme climatic conditions, herders have no incentive to invest in the land.

The transfer of research results is equally as important as their development. An essential element for the successful transfer of technology is the existence of an effective extension system to provide technical assistance. The extension services available to the rural poor are often limited because of the lack of sufficient and qualified personnel, and inadequate knowledge of livestock nutrition, fodder production, use of by-products and potential feed additives. The effective delivery and subsequent uptake of research results, therefore, should be based on a clear extension policy, adequate extension infrastructure and relevant extension messages.

Fodder Planting

Numerous livestock projects have identified the poor quality and insufficient quantity of pastures, forages and crop residues as major constraints to increasing productivity. A significant number have also reported the inadequate use of sown forages and forage legumes, due to lack of farmer awareness and the low availability of forage seed.

Feeding systems practised by the rural poor are mainly based on the use of natural pastures during the rainy season and agricultural by-products and low-cost feed during the dry season. Agricultural by-products such as crop residues, although widely used as sources of feed by smallholders, do not provide the nutrients needed to ensure productive gains. Natural pastures, in addition to having a low carrying capacity, are often inappropriately used and overgrazed. The mismanagement of natural pastures has resulted in soil degradation, an acceleration of erosion and desertification processes, and hence low livestock production and income insecurity.

In most small-scale farming systems, livestock mainly graze on native pastures. and cultivated forages, including cereal and legume residues and, increasingly, fodder trees and shrubs. However, the nutritive value of forages varies seasonally, with significant qualitative and quantitative decreases during the dry season. Smallholders therefore need to grow their own fodder in order to increase animal productivity and reduce the incidence of nutritional deficiencies.

A more practical approach to the problem of feed scarcity is for farmers to establish fodder banks. Fodder banks are plantings of high-quality fodder species. They can be used all year, but are designed to bridge forage scarcity in dry seasons. Fodder banks do not meet all feed needs, but supplement available dry-season forage.

The integration of livestock within cropping systems is becoming increasingly widespread, especially in densely populated areas where land is scarce. This has given rise to integrated cropping systems combining food crop, cash crop and fodder crop production aimed at meeting both livestock and human needs. Such systems have proved to be an effective way to encourage rural farmers to grow animal feed. They are especially popular because of their added benefits, such as manure, animal feed, food and income security, and environmental and soil conservation.

Use of Improved Feed

The growing human population and livestock numbers, decreasing pasture availability and the increasing use of grazing land for crop production have made the use of improved feed in animal production systems all the more urgent. Adequate quantities of high-quality feed are necessary for profitable livestock production. Feeding systems based on improved feeding practices will help maintain the body conditions of animals, increase their resistance to disease and nutrition-related problems, and hence increase animal productivity and rural income opportunities.

The use of improved animal feed is becoming more prevalent, with feeding systems based on improved crop residues and agro-industrial by-products such as molasses, processed cassava and potatoes, oil cakes and milling by-products. Because of feed insufficiencies, animals reared by smallholders often exhibit high mortality, high susceptibility to diseases, low calving periods, late maturity, infertility and reduced reproductive parameters – all of which have obvious negative impacts on household food security and income. Poor feeding practices are the primary reasons for low animal production in developing countries.

The factors that hinder smallholder access to improved feed and the effective use of the vast unexplored feed resources include:

— lack of sufficient land; and where it does exist, it is often unproductive and of low quality;

— limited capital resources, forcing smallholders to seek less costly alternatives that are low in nutritional content; and

— insufficient knowledge of available feed resources, potential feed and its use.

The combined effects of drought and increasing pressure on land have led to major changes in traditional livestock feeding systems. Consequently, feed improvement and use in developing countries has been emphasised in recent years for the following reasons:

— As the human population grows, the demand for animal products increases, thus requiring better animal nutrition based on improved feeding systems. Increases in livestock numbers have boosted the demand for animal feed.

— To feed the increasing human population, more land will need to be devoted to crops, thereby reducing the land available for pasture and fodder. The limited land available for feed and fodder production and the decreasing quality of available pasture have given rise to the use of improved crop residues and by-products as animal feed.

— Increasing human demand for food makes feed improvement essential in order to avoid competition between animal feed and human feed requirements.

In areas of land scarcity, smallholder and landless farmers are faced with the dilemma of choosing between the production of feed for animals and the cultivation of food and cash crops for household consumption and income-generation. Due to limited land, smallholders rarely planted forage crops and relied on crop residues and grass collected from roadsides to feed their cattle. The increase in population and demand for food has resulted in the conversion of grazing land into cultivated fields and its use for other economic activities. Insecure grazing and user rights and uncertain land ownership further aggravate the problem of insufficient land and give smallholders little incentive to invest in feed cultivation.

Government livestock development programmes in developing countries often give low priority to the improvement of feed mainly due to budgetary constraints. Furthermore, most smallholders lack access to improved feed such as concentrates and supplemental feed because of their high costs.

Feed improvement and use in developing countries should therefore be based on:

— farmer practices, production systems and participation;

— on-farm trials;

— locally available and potential feed resources and livestock production systems;

— improved techniques appropriate to the farming system in use and economically and socially acceptable.

References

Aboud, A.A.O., Owen, E., Reed, J.D. and McAllan, A.B., "Feeding sorghum stover to Ethiopian sheep: effect of stover variety and amount offered on growth, intake and selection", *Animal Production,* 1990.

Alvarez, F.J. and Preston, T.R., "Studies on urea utilization in sugar cane diets: effect of level", *Tropical Animal Production,* 1976.

Atta-Krah, A.N., "Fodder trees and shrubs in tropical Africa: importance, availability and patterns of utilization", In: *Integration of livestock with crops in response to increasing population pressure on available resources,* (Editors: T R Preston, M Rosales and H Osorio), CTA: Ede, Netherlands, 1991.

Barry, T.N. and Blaney, B.R., "Secondary compounds of forages", In: *The nutrition of herbivores,* (Editors: Hacker, J.B. and Ternouth, J.H.). Academic Press. Australia, 1987.

Whyte, R.D., *The use and misuse of trees and shrubs as fodder,* Commonwealth Agricultural Bureau. Publication No.10., 1947.

4

Industrial Production of Livestocks

The industrial production system is open both in physical and economic terms. It depends on outside supply of feed, energy and other inputs. Technology, capital and infrastructure requirements are based on large economies of scale and, because of this, production efficiency is high in terms of output per unit of feed or per man-hour, less so when measured in terms of energy units. Yet as the world's main provider of eggs, poultry meat and pork at competitive prices, it meets most of the escalating demands for low cost animal products in rapidly growing urban centres of the developing world.

Industrial production of pork, poultry and (feedlot) beef and mutton is the fastest growing form of animal production. In 1996, it provided more than half the global pork and poultry (broiler) production and 10 percent of the beef and mutton production. This represented 43 percent of the total global meat production, up from 37 percent in 1991-93. Moreover, it provided more than two-thirds of the global egg supply. Geographically, the industrialised countries dominate industrial pig and poultry production accounting for 52 percent of the global industrial pork production and 58 percent of the poultry production. Asia contributes 31 percent of the world's pork production. Industrial ruminant production is concentrated in Eastern Europe, the ex-Soviet Union and in the OECD countries. Typical examples are large-scale feedlots in the USA and in the formally centrally planned economies. Industrial sheep feedlots are found in the Near East, North Africa and the USA.

Environmental Challenges

Because of its open nature and many interfaces with the natural resource base, the industrial 'bio-industry' system signifies for many the epitome of what is wrong with animal production. The industrial scale implies large herd/flock sizes, large volumes of waste, high animal health risks, and often less attention to animal welfare. It has multiple opportunities to dump its waste products without accounting for the environmental costs. The biggest problem, however, is the over-concentration of animals in areas of high human population density with little viable opportunities to utilise waste products on land. The biggest challenge is to establish policies and identify technologies that will help in bringing animal waste in line with the assimilative capacities of land.

Status Quo of Interactions

The industrial system acts directly on land, water, air and biodiversity through the emission of animal waste, use of fossil fuels and substitution of animal genetic resources. In addition, it affects the golbal land base indirectly, through its effects on the arable land to satisfy its feed concentrate requirements. Ammonia emissions from manure storage and application lead to localised acid rain and ailing forests, for example in European countries. Also, the industrial system requires the use of uniform animals of similar genetic composition. This contributes to within-breed erosion of domestic animal diversity.

Animal Waste

Land, water and air are the environmental components mostly affected by the concentration of animals and waste production. Manure is the main agent having effect, mostly during storage and after application on the land. Pigs and poultry excrete some 65 to 70 percent, respectively, of their nitrogen and phosphate intake. Nitrogen, under aerobic conditions, can evaporate in the form of ammonia with toxic, eutrophic and acidifying effects on eco-systems.

A greenhouse gas, nitrous oxide (N_2O) is formed as part of the denitrification process with particularly harmful effects on the environment. Nitrates are leached into the groundwater posing human health hazards, and runoff and leaching of nitrogen directly lead to eutrophication and bio-diversity loss in surface waters and connected ecosystems. Phosphorus, on the other hand, is rather stable in the soil, but, when P saturation is reached after long-term high level application of manure, leaching occurs and this also causes eutrophication.

Ammonia and other nitrogenous gases result from the digestion of protein, part of which is lost in manure and urine. Growing pigs, for example, excrete 70 percent of the protein in feed while beef cattle excrete 80 to 90 percent and broiler chickens 55 percent. Ammonia, in high concentrations in the air, can have a direct effect on plant growth, by damaging leaf absorption capacities but its indirect effect on soil chemistry is even more important.

Ammonia acidifies the soil, interferes with the absorption of other essential plant elements, particularly in nitrogen-poor ecosystems such as forests. Livestock production is a major source of ammonia emissions in the industrial world. For example, of the 208,000 tons of ammonia emitted in the Netherlands in 1993, 181,000 tons was estimated to come from manure. This was about 55 percent of the total acid deposits in the Netherlands, industry and traffic being other important contributors.

The various forms of nitrogen losses lead to much reduced levels available for crop nutrition. Plant uptake and use depend on a series of other factors such as species, climatic and soil conditions. According to Bos and de Wit, 20, 50 and 44 percent of the nitrogen excreted by pigs, broilers and laying hens are lost to the atmosphere as NH3. Because of its different chemical properties, phosphorus losses are insignificant because phosphorus is not very mobile in the soil.

The amount of N, P, K and other nutrients available to the crop within the soil determine the fertiliser value of manure. Further significant losses may occur depending on the type of stable and manure management system and thus define the direct environmental impact. Substantial nitrogen and phosphorus losses also occur when manure is applied on the land. The spreading of manure directly on the land can lead to nitrogen leaching into water as nitrates and contamination of surface waters. This in turn leads to high algae growth, eutrophication and, as a result damages aquatic ecosystems. Not all soils are equally susceptible to nutrient loading and contamination. Sandy soils with low cation exchange capacity and, consequently, poor retention characteristics and high runoff, are particularly at risk.

Heavy Metals

Copper and zinc, which are essential minerals in animal nutrition, are deliberately added to concentrate feed whereas other heavy metals, in particular cadmium, are introduced via feed phosphates. Only 5 to 15 percent of metal additives are absorbed by animals, the rest is excreted. Soils, on which poultry and pig manure are continuously applied at high rates, accumulate heavy metals, jeopardising the good functioning of soil, contaminating crops and posing human health risks.

Fossil Fuels

The industrial system is a poor converter of fossil energy. Fossil energy is a major input of intensive livestock production systems, mainly indirectly for the production of feed. Brand and Melman show in case studies that, typically, feed accounts for 70 to 75 percent of the total energy input except for veal production where it is almost 90 percent. Energy output for livestock products comprises food and non-food items. Southwell and Rothwell calculated output/input ratios of 0. 38, 0. 11 and 0. 32 for pork, poultry and eggs respectively, considering fossil energy only.

A large portion of non-food energy output is in the form of manure and the potential for recovery of this energy in the form of methane has greatly increased in recent years. The heavy concentration of animals in certain regions, particularly of pigs and poultry, has given rise to the development of large-scale processing for use elsewhere. Most problems lie in high energy expenditure for drying and for transport.

Biodiversity

The industrial system has a threefold effect on species wealth through:

— Its demand for concentrate feed , which changes land use and intensifies cropping. The production of feed grains, in particular, adds additional stress on biodiversity through habitat loss and damages in ecosystem functioning.

— Waste production and its effects on terrestrial and aquatic ecosystems. These effects are often geographically confined to areas of high livestock densities. Eutrophication and destruction of habitats is a common phenomenon in parts of north eastern Europe and the USA as well as in the densely populated areas of the developing world, in particular Asia, and to a lesser extent, Latin America. Ammonia emissions lead to acidification of particularly fragile habitats and so causes losses in biodiversity.

— The requirement for extremely uniform animals of similar genetic composition contributes to within-breed erosion of domestic animal diversity.

Benefits of Industrial Production Systems

First, the rapid development of 'modern' industrial pig and poultry systems helps to reduce the total feed requirements of the global livestock sector to meet a given demand. It may therefore alleviate pressures for deforestation and degradation of rangelands, such as is happening in parts of

Latin America and Asia, thus saving land and preserving biodiversity. Second, the feed-saving technologies developed for this system can be effective at any scale and therefore can be successfully transferred to mixed farming systems. The same holds true for waste prevention and treatment technologies which have been developed following regulations applied mainly to the industrial system. There, the resource-saving and waste management technologies generated by the industrial systems bring benefits to the sector as a whole.

Driving Forces in Industrial Production

Population growth, rising income and urbanisation are the fundamental driving forces determining growth of industrial livestock production. Globally, industrial animal production is the fastest growing sector, with over 4 and 5 percent growth per year in pork and broiler production, respectively. The Annual growth for eggs is 3.8 percent and for mutton and beef 2.5 percent. Driven by rising incomes and rapid urbanisation, Asia experienced a staggering growth of 9 percent per year in industrial pig and poultry production over the last decade, trailed by Latin America with a strong growth in poultry.

As shown above, it is not industrial production per se which creates environmental problems but the fact that production units tend to concentrate in certain areas. The reasons are the following:

— Transport costs and market opportunities are the main determinants. The industrial system. especially for pigs and poultry, is characterised by a significant, and often exclusive, use of feeds of high energy content (mainly cereals, oilseeds and their by-products). This high energy density allows movement of feeds over longer distances than perishable animal products, even though the quantities are larger. These relative costs encourage production, slaughter and meat processing facilities to be located near urban centres. The large number of industrial units being established around Beijing,

Shanghai, Mumbai and Calcutta demonstrate this incentive. Furthermore, in net grain importing developed countries, with abundant road and cooling infrastructure, large-scale industrial operations are located close to ports, such as pig operations in the Netherlands and northern Germany.

— The agrarian structure has favoured high livestock densities and the evolution of industrial-type systems where shrinking farm size forced farmers to engage in value-adding activities without sufficient land available, such as in some southern parts of Germany and the Netherlands, or to abandon agriculture altogether. This move has been favoured by price subsidies with the main objective of supporting rural populations.

— Different policy settings within a country or free-market-zone through, for example, tax advantages or low environmental standards, may lead to concentration of the industrial system. An example is where intensive poultry, and more recently also pig production moved, from the corn belt of the USA to the southern states and, in the EU, to Italy and Spain.

In the past, industrial and intensive mixed farming systems have benefited from policy distortions and the absence of regulations or their enforcement and, in many cases, this vacuum has given this system a competitive edge over land-based systems. Furthermore, some policies have misdirected resource use and encouraged the development of technologies which are inefficient outside the distorted context. In the EU, for example, high domestic prices for beef, pork and milk have benefited industrial production.

In the Near East, small ruminant feedlots are heavily dependent on subsidised feed, and this has encouraged inefficient feed use. Likewise, in the former centrally planned economies, feedlots were based on heavily subsidised feed grain and on subsidised fuel and transport. In many developing countries, there are not only direct subsidies on feed but also on energy. As energy is a major direct and indirect cost item in industrial production systems, economy-wide policies often tend to favour industrial production over grazing systems and mixed farming.

Finally, in practically no country in the world, is the industrial system charged with the full environmental costs of production. It appears that societies prefer the cheap supply of animal products over the functions of the ecosystems concerned. Self-sufficiency in animal products and supply of high-value food commodities to urban populations seem to be overriding policy objectives, particularly in developing countries.

Impact of Appropriate Technologies and Policies

In the developed world, the pollution of land, water and air has raised acute awareness of the environmental problems associated with industrial livestock production. This has, in many cases, triggered the establishment of policies and regulatory measures that address these problems. To the contrary, the absence of regulations and their enforcement, together with a surge in demand and a continued indifference about growing environmental hazards in the developing countries call for immediate action. The challenge is to identify and mix the appropriate technologies and policies that suit local circumstances.

Policies and Regulations

— Regulatory instruments are imposed to control the distribution and concentration of livestock production and introduce technical control systems. Some of these regulations are relatively easy to enforce whereas others, for example the maximum permitted amount of manure per unit area, are more difficult. The regulatory approach is most efficient in situations of point source pollutions, i.e where the polluter can be unmistakably identified and where there are strong institutions. In countries with weak institutions, the enforcement of regulations at reasonable social cost remain a major challenge and limits the validity of this approach. Compliance with regulations affects the cost of production and may therefore influence regional distribution. Examples include limits on the number of animals in most of the member states of the EU, taxes on surplus animals as in Belgium, taxes on surplus P (most EU countries), a ban on direct discharge of manure into surface waters (USA, Malaysia) and the establishment of nutrient management plans (Indonesia, USA, Europe). Guidelines on manure storage and application methods, timing, crops and quantities are available in practically all countries with high animal densities.

— Zoning can regulate regional distribution. Zoning is an important policy for controlling animal manure storage and processing, not only for environmental reasons but also for concerns of human health and rational regional development. Zoning has been important both in environmental and in regional development policies, as well as in successfully moving industrial production units away from urban centres in OECD countries. An important prerequisite for successful zoning is good infrastructure because animal products will have to be transported over larger distances.

Marketing and processing infrastructure must therefore be taken into account when defining zones in order to make the best use of investment. The creation of confined industrial 'parks', with prescribed and sometimes shared facilities for waste collection and treatment, offers opportunities to fully charge industrial production systems with environmental costs while still maintaining advantages of market access and economies of scale. Governments have frequently established guidelines for the siting of production units, particularly to protect urban settlements from obnoxious smells. Zoning, when done within a comprehensive area development plan, also allows for common waste collection and treatment facilities to be shared by a number of producers.

A trend to rational zoning is not only fostered through environmental concerns but also by changes in overall policies, often triggered by the removal of government interventions and trade liberalisation. In the Near East, for example, industrial small ruminant systems have been kept viable through subsidies on grains and are increasingly under pressure because of the financial requirements to maintain these subsidies.

— The most efficient and direct financial instrument would be to internalise all environmental costs into the consumer price. However, implementation of such policies is not easy. First, there is a lack of accurate economic evaluation of these costs, for example for biodiversity and some gaseous emissions and some indirect costs (soil erosion of feed production, for example). Initial calculations point to an increase of 10 to 15 percent in cost price. Second, unequal application of the inclusion of environmental costs in the product price puts some producers at a disadvantage.

Current financial instruments therefore focus on reducing emission of nitrogen and phosphates and other potential pollutants, particularly in susceptible and already burdened areas. The levies and taxes currently imposed on the intensive mixed and industrial systems in pratically all developed countries fall into this category. Others are:

— Removal of subsidies on concentrates to increase the cost of feed concentrate—intensive production and to favour land-based systems over the industrial system. In a more indirect way, removal of subsidies on, or taxation of, fossil fuel may have a similar effect by raising the cost of feed. Government income from such sources could be used to alleviate environmental problems.

- Removal of import restrictions on materials and equipment that improve feed efficiency, such as amino-acids to improve protein, enzymes to improve phosphate digestion and feeding equipment that restrict intake. These technologies lead, indirectly, to lower waste loads through better feed conversion.
- Subsidies for investment or running costs to improve the adoption of emission control technologies. For example, subsidies for constructing manure storage facilities are given in many EU countries. In the USA, cost-sharing and state revolving funds have been established for manure storage sheds and dead bird composters.
- A system of tradable manure emission quota would limit waste production and still create incentives for efficient resource use. Marketable permits and pollution trading is based on the establishment of payment per unit of pollution or the use of pollution reduction credits.

New Techniques

The above policies have been accompanied by the introduction of a wide range of new techniques. Because of the commercial and demand-driven nature of the industrial system, development and transfer of technologies are usually not a problem. A whole range of technologies exists that could alleviate the environmental burden created by this system and seek improvement in two areas:

- Reduction of nitrogen and phosphate excretion by improving feed utilisation can be achieved through:
    - Introduction of multi-phase feeding in order to match feed composition to the needs of the individual animal classes. By better adjusting the nutrient supply to the needs of animals, less waste is produced and therefore less nitrogen and phosphates are released in waste and into the environment.

  - Improving the accuracy of determining nitrogen and phosphate requirements, followed by better balancing of feeds with these essential nutrients. In this respect, important gains have already been obtained in better balancing pig and poultry rations with essential amino-acids, the building blocks of feed proteins.
  - Increasing diet digestibility. Spectacular improvements have been obtained with the addition of an enzyme (phytase) for catalysing the digesting of phosphates in feed. The same enzyme might also increase the availability of zinc in feed, reducing the need for feed additives;
  - Promoting feeding systems which reduce intake and stop the buffet-style, ad libitum feeding which was popular in the 1980s.
- Reduce the emission from manure storage and during application. Nutrient losses from manure in stables and during storage can be reduced through improved collection and storage techniques. In animal buildings, manure is stored either under a solid floor, under a slatted floor, or within a housing system with litter. A large part of ammonia emissions comes from the manure surface in storage, either under the slatted floor or in open tanks. The main focus has to be on reducing nitrogen losses, most of which are in the form of ammonia from the manure surface. Possibilities are:
  - solid storage systems which collect the faeces only for storage in bulk, and have low gaseous emissions;
  - liquid or slurry systems where animals are kept on slatted floors and the manure and urine are collected and stored in large tanks;
  - lagoon systems consist of flushing systems using large amounts of water. Nitrogen losses and methane emissions are high;

— covered tanks and manure kept under solid floors in stables significantly reduce emission problems. Reductions in emissions of 80 to 90 percent can be achieved by covering storage tanks.

Methods that efficiently reuse energy and nutrients in manure in cases where manure is not directly used for agriculture. Biogas plants of all sizes and different levels of sophistication exist. They not only recover the energy contained in manure but also eliminate most of the animal and human health problems associated with micro-biological contamination by microorganisms.

Other methods of controlling the waste load are purifying and drying the manure; reducing nutrient losses during and after application of manure on soils. Injection or swad application of manure into the subsoil and appropriate timing significantly reduces losses. Nitrification inhibitors can be added to the slurry to decrease leaching from the soil under wet conditions.

Large-scale manure processing is possible where intensive production is concentrated in certain zones but that is often not viable economically. The efficient use of manure for feed and energy production requires high capital investments which often cannot be borne by individual farmers.

Recycling manure by feeding it back to livestock, including fish is practised only on a limited scale. In addition to widespread reluctance to use manure as feed, mainly originating from fear of health hazards, most types of manure have a low nutritive value with the exception of poultry manure which is of a reasonable quality. In intensive production systems with large amounts of collectable manure, cheaper feeds are also available, while in production systems where utilisation of low quality feeds is common, high collection and opportunity costs (manure as fertiliser or fuel) prohibit the use of manure as feed.

References

Slaybaugh-Mitchell, Tracy L., *Indigenous livestock production: An annotated bibliography of husbandry practices (Bibliographies in technology and social change series)*, 1995

Umali, D., Feder, G. & de Haan, C., "The balance between public and private sector activities in the delivery of livestock services", *World Bank Discussion Papers No. 63*, Washington, DC, World Bank, 1992.

5

Urban Livestock Keeping

Livestock raising has often been part and parcel of urban agriculture, presenting its own specific problems and opportunities. Until recently, it was often regarded as problematic, backward and a sign of poverty. As with all branches of urban agriculture, however, livestock keeping now seems to be recognized for the positive role that it can play in urban living conditions across the world. Indeed, livestock production has a variable and controversial, but often essential, role to play in and for cities. It occurs on a small scale, with both small and large animals. It is subject to change, e.g. local breeds are replaced by foreign breeds and, while it is increasing in some cities, it is decreasing elsewhere.

Stakeholders in Urban Livestock Keeping

Among the stakeholders in urban livestock systems are the shantytown women and children who make a living by, among other activities, feeding their food wastes to rabbits, guinea pigs or poultry. Other stakeholders are government officials, entrepreneurs, shopkeepers, leaders of dairy cooperatives, feed and medicine merchants, retailers and wealthy consumers. Examples of various urban livestock activities and the occupational groups of the people involved are given in Table 1.

The differences among stakeholders imply different interests and perceptions regarding urban livestock. Government officials may know little or nothing about the socioeconomic and technical aspects of animal production but they can be zealous in protecting their own private interests. Entrepreneurs may be retired or active government employees, their wives, single women or unemployed casual labourers; they may be schoolchildren or industrialists who know how to make good use of government subsidies and tax laws. Combined with the different forms of urban livestock, the result is a wide range of different ways of perceiving problems and opportunities in urban livestock; and some form of categorisation is required to cope with this variation.

Table 1. Non-agricultural and agriculture-related occupations of poultry producers in the United Republic of Tanzania

	Dar-es-Salaam		Mwanza	
	(No.)	(%)	(No.)	(%)
Occupation				
Shopkeeper/trader	8	36	5	15
Retired/unemployed	3	14	4	12
Professional/office employee	1	5	14	41
Sewer/tailor	1	5	8	24
Military/government employee	0	0	2	6
Other occupation	9	39	1	3
Activity				
Keeps dairy cows	23	37	18	50
Sells livestock products	6	10	1	3
Operates feed/maize mill	2	3	3	8
Farms	25	40	4	11
Involved in veterinary or agricultural extension work	2	3	3	8
Keeps small livestock	3	5	4	11
Other agricultural occupation	1	2	3	8
Total	62	100	36	100

Note: Some producers gave more than one response.

For instance, all those people who own a bullock cart could be regarded as representing one group; other groups could be all those who own two, or three or more dairy cows. Such people depend on their livestock for at least part of their and their families' livelihoods, and they will be reluctant to give up this source. The same is true of single parents who are raising extra income, improving their children's diets or earning money for school expenses by keeping animals, which also help to dispose of kitchen waste or by-products from the local industry.

However, not everybody within the same level of urban society shares the same positive view of urban livestock keeping. For example, a livestock keeper may live next door to the mother of a child who comes home with clothes dirtied by animal excreta, or there may be a neighbour who gets angry about goats causing havoc in his vegetable plot.

A second category of people consists of local legislators and civil servants. Their main concerns are about disputes in their neighbourhoods, or about raising some extra money by fining people for breaking the law. Many administrators regard urban livestock keeping as a sign of backwardness, but there are exceptions. For example, much of the urban dairy business of Dar-es-Salaam in the United Republic of Tanzania is carried out by the city's civil servants as a sideline.

A third category of stakeholders in urban livestock are those administrators, and often academics, who are professionally geared to see only one aspect of what ever it is they are looking at. For example, people who work in the public health office will be concerned only about public health; tax collectors will worry only about tax (and fee) collecting; those in charge of urban housing will concentrate on construction plans and ignore the animals; and veterinarians will be interested in zoonoses and private practice opportunities. This description may be somewhat exaggerated but it does characterize the ways in which many employees of government offices tend to function.

A health official cannot compromise on public health even when animals represent an essential form of income that helps local people to survive; a tax-collector cannot tolerate informal arrangements that are difficult to administer, even when they form an essential part of local livelihoods. Reports from such sector departments and/or discipline-oriented professionals are very likely to emphasize problems in whatever area they are regulating or studying. The identification of creative and holistic solutions is not part of their mandate, nor have they been trained to do so. One possible way of improving livestock systems, therefore, lies in the (re-)training of government officials to use their rules to help strengthen the positive aspects of urban livestock while overcoming the negative ones.

A fourth category of stakeholders in urban agriculture is that of the national and international planners and policy-makers who are concerned with the production of sufficient food, of plant and animal origin, to ensure that people get enough to eat. These people tend to emphasize the fact that urban livestock and agriculture produce only a fraction of the dietary food requirements for an urban population, and they rightly concede that urban livestock and crops can never produce enough.

A combination of the possible pollution and health risks of keeping animals in close proximity to people and the competition with other sectors (for space and resources), leads planners and policy- makers to suggest that the production of food from animal origin is better shifted to larger specialised units outside the city. As with the discipline- and sector-oriented people, these people also tend to overlook the fact that urban livestock can fulfil many roles besides producing milk or eggs, and that a small contribution to food security can make a lot of difference at the family level. Among planners and policy-makers it would be beneficial to increase awareness about the essential roles that animals can play in the economy, sociology and ecology of cities.

A fifth but very relevant category cuts across all the others. This categorisation distinguishes between, on one hand, so-called "linear" thinkers who are interested in only one aspect, e.g. food production and, on the other hand, "non-linear" thinkers who consider several aspects and interests to reach holistic approaches to the issue. The two lines of thought should supplement each other, but in practice they tend to run counter to each other. Successful action is often the result of local creative initiative rather than general across-the-board legislation.

Major Systems In Urban Livestock Keeping

Urban livestock systems can be defined in many different ways and at many different levels. A global-level classification, which may be useful for organisations such as FAO, may be completely irrelevant for interventions at the city or suburb level. At that level, for example, a distinction may be made between monogastric and ruminant systems (pigs, chickens and ducks versus cows, buffaloes and goats or sheep), but that may, in turn, be irrelevant when two neighbours quarrel about their respective systems being a public nuisance or an asset.

Participation of local stakeholders is indispensable to choose relevant criteria and discard irrelevant ones. However, the major patterns that occur in urban livestock systems seem to follow a set of classification criteria based on differences among cities, population densities and histories, local stakeholders, animal species, etc.

All cities differ from one another in size, development history, sociology and structure. Mexico City has a population of around 22 million people, while Abidjan (Côte d'Ivoire) and Dar-es-Salaam (United Republic of Tanzania) count about 2 and 3 million people, respect ively. When thinking in terms of future developments in opportunities for urban livestock, distinctions can also be made between cities that have grown in blanket form and cities that were planned according to the concept of satellite towns. Obviously, a range of intermediate forms exists between blanket and satellite patterns, and practice does not always follow planning.

Some cities opt for decentralised planning or laissez faire, others for centralized planning, and still others seem to have developed without any clear plans at all. Singapore is a typical case of a city with strong legislative measures, but in other cities hardly any effect is expected from legislation other than an increase in bribery. Keeping dairy cows may be impossible in the heart of modern Mumbai (India), Tokyo (Japan) or Amsterdam (the Netherlands); but rabbits, chickens and other small (pet) animals are commonly kept on balconies and in backyards everywhere.

Even the keeping of dairy cows and/or goats can be quite acceptable in urban areas with much green, or where there are a lot of by-products from agro-industries. In the nineteenth century, urban dairies were quite important in areas around beer breweries in Copenhagen (Denmark) and around distilleries in major United States cities, while today they still supply a good income in Dar-es-Salaam, particularly to civil servants.

Another potential criterion for distinguishing among urban livestock systems is in their origins. A system may emerge through:

— "organic" emergence, i.e. the system evolved by itself;
— induced development, often with public money;
— city expansion, e.g. when previously existing village farms are swallowed up by urban development to create overgrown farming;
— importation of livestock by rural people immigrating into the city;
— traditional tribal groupings that settle in towns while maintaining their liking for animals;
— situations in which livestock is seen as an emerging opportunity for cleaning waste (Montevideo [Uruguay], Ghana)
— cities with wealthy people where animals can provide emotional attachment or a learning opportunity.

The terms "organic emergence", "induced development", "overgrown farming" and "imported by immigration" are useful classifications, particularly when governments have to negotiate with stakeholders in urban livestock systems about future courses of action.

Scale and Degree of Commercialization

The scale and degree of commercialisation are perhaps the most important criteria for distinction among urban livestock systems. Basically, distinctions can be made among subsistence backyard, semi-commercial and large-scale commercial systems. Large-scale commercial systems may ultimately be the most problematic ones to maintain. They often have few roots in the social network and tend to be associated with large amounts of waste such as dung and urine; they also tend to concentrate the social and economic benefits in the hands of a few and to disperse the costs across society.

In these ways, large-scale commercial enterprises much resemble their "role-models" in industrialised systems, and they are now restricted or banned in Western countries and, for example, in Singapore. Smaller-scale subsistence and semi-commercial systems are based more on local recycling of nutrients; they allow local control over excessive pollution and they tend to internalise their own problems.

Species and Animal Types

The types of animals kept in urban livestock systems range from dairy cattle and buffaloes, via rabbits, goats and guinea pigs, to backyard poultry, industrially kept layers and broilers, and pigs of native and exotic breeds. Large farms as well as small farms can be either clean or dirty; animals kept in small cages can be clean and well maintained, although this is currently attracting much criticism from animal welfare groups. Animals can be kept in clean sheds, and excreta can be disposed of in a biogas pit or made into dungcakes for sale or use in the home. Ethnic and religious beliefs make for remarkable differences in preference for animal species: pigs are popular in China, goats are often kept in Moslem countries, and Hindu societies have a preference for cows.

Degree of Integration with other Systems

An urban livestock system can consist of anything from one

woman with one goat and/or some chickens in the backyard to a large company with more than 30 000 chickens. Another example of an urban livestock system is the Landhi milk colony in Karachi (Pakistan) where tens of thousands of cattle are kept in one place, most of them individually owned. This production pattern is a highly specialised, purely business-oriented system that fully exploits the neighbouring high-demand market, as well as the availability of buffaloes that have recently calved and of feed and fodder sources from the Sindh and Punjab provinces.

In India, large dairy enterprises in, for example, Mumbai and Pune purchase buffalo and dairy cow calves from Haryana and western Uttar Pradesh, a distance of about 1,000 km by road. Similarly, in Abidjan (Côte d'Ivoire), beef comes from the grasing areas far north; in Damascus (Syrian Arab Republic), day-old chicks come (or used to come) from Europe; and roughage for dairy and beef animals in Khartoum (the Sudan) originates largely from the Gezira scheme, an irrigated area some 100 to 200 km south of the city.

The cobblestones in Pompeii (Italy) bear the scars of wear and tear through intensive cart traffic. In Madrid (Spain) there was a significant peri-urban livestock keeping sector until very recently. Sumberg shows how urban livestock systems have been very common in the United Republic of Tanzania for at least a century. After the major earthquake that hit Mexico City in the 1980s, many dead pigs were found in the rubble - animals that had been kept in apartments often several storeys high.

Across Europe there are towns and cities with traditional "hay" and "straw" markets, and many squares and streets still carry that name. These sites were used for the trading of straw and hay which came from rural areas to feed the horses, oxen and cows that were needed for transport and/or production of milk, etc. in the city. Phelan recounts how dairy cows used to be a common sight in Copenhagen (Denmark), where they cleaned-up leftovers from the beer breweries. Obviously, there was and is a point in keeping livestock in cities, either as a specialized enterprise or in combination with gardening and farming in and around the city. Urban livestock systems exist all over the world, in spite of the many problems they are perceived as creating.

Factors in Urban Livestock Keeping

Many modern urban livestock systems occur in tropical regions, but their existence has been described for a long time, for example by the agricultural economist Von Thünen, who researched German conditions some 150 years ago. In his "location theory" he showed how intensive livestock systems tend to occur right in and around urban centres, together with other intensive forms of agriculture such as vegetable production.

The production of forests for energy and timber took place at a somewhat greater distance from the towns, followed by cropping and more extensive forms of agriculture and, ultimately, by extensive livestock production. Obviously, the location theory should not be taken too literally; fossil fuel has replaced timber and the general pattern may be disturbed by rivers, roads, mountain-ridges, etc. Still, the basic outline of Von Thünen's theory should be kept in mind when discussing the emergence of urban livestock systems in any part of the world.

It can be hypothesized that (peri-)urban agriculture and urban livestock systems originated from several complementary factors, including:

— the availability of a choice of high-quality feeds such as by-products of the food processing industry (bran, oilseed cakes), hotel refuse and kitchen waste. These feeds may be available in a scattered fashion in the countryside, but in the city they can profitably be combined into nutritionally and economically more optimal ratios;

— a strong local demand for food and power from animals, often resulting from the presence of wealthy consumers. Note that a moderately wealthy population tends to consist of a poor and a rich sector. The rich, in particular, verge towards increased consumption of food from animal origin;

— the potential of vegetable and livestock units to provide a high turnover, which gives a relatively high income on small plots of expensive land;

— the presence of significant proportions of poor people and wealthy investors, both of whom find an extra source of food security and/or income through the keeping of animals.

Other factors that affect the tendency for urban livestock systems to emerge include the availability of specialised services such as veterinarians and processing industries and the need for short distances between producers and consumers for commodities such as fresh milk and "hot" meat. These factors combine with the availability of more land at a greater distance from the city, where staple crops such as cereals can be grown together with feed and where animals can be raised for fattening in the city or to serve as pets for urbanites. The upland town of Batu (East Java, Indonesia) used to have a thriving market of pet rabbits, which were sold at weekends to the wealthy upper class from nearby Surabaya and Malang.

Similarly, cattle and sheep fattened around London (the United Kingdom) were (and still are) raised and grown in Scotland; animals raised in rangeland areas of the southern Sahel are finished in cities such as Abidjan; and urban dairies of Karachi and Mumbai transport their lactating animals for long distances. Knowledge and understanding of these factors are essential before ad hoc improvements in urban livestock keeping can be attempted.

Issues and Challenges in Urban Livestock Systems

Many urban livestock systems are undergoing a period of crisis. In many government offices, there is a feeling that urban livestock systems are under pressure from a variety of issues.

Many of these issues are clearly problematic, others are open to several interpretations.

Urban livestock systems as a potential hazard to public health. Potential hazards for public health range from poor hygiene, caused by the presence of dung, flies, parasites and zoonoses, to the difficulty of controlling product quality when food from animal origin is used directly by consumers. For example, livestock keeping without a proper sewage system may favour mosquitoes that transmit malaria and some important viral diseases, such as yellow fever, dengue, Rift Valley fever and West Nile virus.

Zoonoses are diseases that affect both humans and animals. They are more likely to spread when hygienic conditions are poor - for example, commercial forms of livestock keeping in urban areas are particularly favourable to the multiplication of rodents because these systems require the storage of animal foodstuffs. Rats are the reservoir of, for instance, hanta virus, an important human disease in Asia. In Africa and Latin America a human tapeworm makes use of scavenging pigs to complete its lifecycle.

Pigs feeding on human excrement containing tapeworm eggs become infected by cysticerci (the larval stage),while humans who have eaten insufficiently heated pork meat become hosts to the adult tapeworm. The disease becomes a serious problem when other humans, such as playing children, are infected with tapeworm eggs from human excrements. The housing of animals, maintenance of hygiene and proper storage of animal feeds are typical responses to these problems.

Urban livestock systems as a potential hazard to animal health. Poverty, ignorance and a lack of veterinary services favour the spread of human and animal diseases. In addition, where there are poverty and administrative incompetence, these diseases are less likely to be noticed by veterinary officials. In Ghana, the first outbreak of African swine fever (ASF), an endemic disease in East Africa, occurred among backyard animals. The disease had time to become rooted in backyard pigs across Ghana before the cause and significance of the subsequent mortality were discovered by the veterinary services.

Uncontrolled slaughter aggravated the spread of ASF. In a similar way, it may have been refugees from Mozambique who brought Newcastle disease to Zimbabwe some years ago. The severe mortality in chickens all over the country was difficult to constrain, as adequate vaccination campaigns in backyard poultry were complicated to organize.

However, in spite of the thinking prevalent in many government offices, it is incorrect to blame these problems solely on backyard farming. Large losses have been incurred in so-called modern commercial and industrial systems, such as the outbreaks of classical swine fever and bovine spongiform encephalopathy (BSE or mad cow disease) in Western Europe.

Pollution from manure effluent and abattoir waste. Pollution from animal production is caused by a range of issues, some of which are poorly specified and based on conjecture or local observations. Potential problems with dung and urine disposal are obvious; flies breeding on animal and vegetable waste may transmit food-borne diseases, such as diarrhoea. Problems arising from wastewater and rejected meat from abattoirs (slaughterhouses) are less obvious to most of the general public. When the building construction is adequate, there is proper meat inspection and strict removal and destruction of rejected carcasses, abattoirs do not pose a risk for human health.

Depending on the local culture, it is quite common to find very clean processing of animal produce. The pollution issue needs to be balanced with the fact that livestock keeping using the resources available in (peri-)urban areas contributes substantially to the cleaning of local waste. Chickens keep the backyard clean and create value from leftover food, while pigs utilise hotel waste and by-products from agro-industries, thereby reducing the amount of organic waste to be disposed of. Some cities actively promote programmes to recycle poultry waste for agricultural purposes (Ghana).

Only a small proportion of urban food requirements comes from urban livestock. A major concern regarding urban livestock systems is that the (growing) demand for food, particularly for wealthier consumers, from animal origin in urban areas is outpacing the production potential of urban systems. Urban livestock may never be capable of producing enough food for the entire city (even when food production is the only function of livestock), but the urban livestock system's share of food production can be substantial for some sectors.

An important point for understanding urban livestock systems is, however, that animals have many functions. In addition to providing food, they:

- provide social comfort and emotional attachment;
- convert waste;
- can be a source of income;
- serve as a savings account;
- provide energy (draught power).

Indeed, small-scale enterprises keeping rabbits, guinea pigs or even poultry, goats or cows often start up in areas where no similar activities existed previously and generally do not demand public money. For example, rabbit keeping was adopted in London during the Second World War as a solution to issues of food scarcity. The smaller forms of animal production are very dynamic and adaptive. They provide income and useful products to the city's inhabitants and they contribute, even as a side activity, substantially to the livelihood of many households. Generally, even in cities, a rural ("green") landscape is appreciated; this is reflected, for instance, by the higher prices of houses facing farm plots or parks instead of apartment buildings.

Management of Technology and for Livestock Development

The many benefits of urban livestock are counter balanced by a series of problems that can be dealt with in a variety of ways. Complex problems do not have fixed solutions because each solution brings its own problems, depending on the perception of the stakeholder and the boundaries of the system involved. Many methods have been developed to address these issues of complexity.

Most of the methods are based on participatory approaches that take into account the perceptions of the different stakeholders at several levels of the system hierarchy in order to make interpretations, classifications and solutions (coping strategies) acceptable. For example, not only does the local livestock keeper have to agree with the solution, but so also do the administrators (and vice versa).

Livestock keepers can increase the output of their herds by using various technologies and/or management methods. Local communities can be instrumental in solving their own problems, as long as administrators are willing to be flexible and creative - an approach that is likely to be very useful in overcoming problems and constraints in urban livestock keeping.

A strict distinction between technology and management cannot be made, but it should be clear that proper management can save resources. This is important because farmers are concerned with costs as well as outputs. The issue becomes more complicated when different farmers attribute different values to inputs and outputs. For example, a resource-poor farmer (RPF) who is short of cash may want to realise maximum returns on his or her cash and aim at minimising investments, resulting in low costs but also low benefits.

On the other hand, a resource-rich farmer (RRF), who has plenty of cash at hand, might pay for extra labour or opt to use commercial feed or expensive housing, thereby increasing production but also costs. Comparing these two categories, the RRF is likely to have a higher absolute margin (total benefits minus total costs) than the RPF, but the return on cash (benefits related to investments) of the RPF is likely to be better.

A high output does not always guarantee a high monetary return. Moreover, certain feeds (i.e. resources) may have unexpected qualities at unexpected occasions. A typical example of this is the cost of feed energy from straw and concentrates; straw is the cheapest source of nutrients in village conditions, but the most expensive in towns. However, in spite of the high costs, concentrates are fed in villages because animals cannot eat enough straw to support high levels of production, and straw is fed in cities because it provides fibre for rumen function which is critically lacking from concentrate-rich feed rations.

Last but not least, a farmer may be overfeeding the animal, making it overweight and leading to fertility problems or other health complications; a rabbit owner may build a hutch that is too protective (no fresh air) or too expansive; a family may make large investments in a biogas pit that turns out to be useless because the liquid flow is not well maintained.

Technology needs to be well managed and proper management needs the right technology. Not every technology is useful, and it is important to select the best technology for each individual case. To make matters more complicated, any technology and/or management method is associated with several costs and several benefits, some of which are difficult to measure, and some of which are measured quite differently by, for instance, farmers and policy-makers.

Some costs and benefits become visible only after some time. For example, the apparent advantages of agrochemicals turn out, in time, to have negative trade-offs such as reduced disease resistance and increased toxic residues. The emphasis on high input levels in urban industrial livestock systems has led to the decline of those systems. Large surpluses of dung and depressed prices have been among the factors leading to this demise.

Farmers and/or animal keepers in the subsistence and semi-commercial sector work on a small scale, quite often with a high labour input and in close contact with immediate neighbours and consumers. They purchase few inputs and use kitchen or market leftovers. Basically, these farmers adjust their products to the available resources. Large-scale farmers adapt the resource flow to their own industrial targets, and large (infrastructural) investments make them rather rigid. Their advantages are that they generally have more political influence and access to external information. In the long term, the future of large-scale farms gives more cause for concern than that of smaller-scale ones.

Classification of Technologies

As with farmers, there are many different types of technologies. This section makes distinctions among different technologies related to urban livestock systems in general. More specific technologies oriented to different livestock types (focusing mainly on backyard animals), disciplines (feeding, reproduction and animal health) and sectors (processing, waste management).

Input- and Management-based Technologies

Input- and mechanisation-oriented technologies increase the output of a particular animal or farm by adding more inputs (in the city, these are feed, medicine, etc.) or by using a machine to save labour or time. Management-based technologies imply that a farmer solves problems, not by using more inputs, but by being very careful as to where and how existing inputs are applied. For example, when it first stands up in the morning, a cow takes some time to stretch, urinate and defecate. The farmer therefore waits for 15 minutes before letting the cow out, so that the dung and urine can be collected from the animal shed, which saves on fertiliser use. Proper management can also help to improve heat detection and reduce disease because farmers watch their animals more carefully. Manage ment can lead to decisions to change the number of animals or the herd composition in order to make better use of existing resource flows.

Accelerating and Defusing Technologies

Accelerating technologies tend to treat symptoms with readily available, "linear" solutions. Instead of solving the problem, this can lead to an aggravation of the problem. In contrast, defusing technologies look, in the first place, for the underlying causes of the perceived problem and come up with alternative ways of dealing with it.

The following are examples of accelerating versus defusing technologies in urban livestock keeping:

— Using medicines to force a cow to conceive when its condition does not allow conception may result in the collapse of the cow. On the other hand, if the cow is dried for a few weeks it has more chance to build up the reserves necessary for the next gestation and lactation. Another option would be to return the cow to a village where she is able to recuperate and can conceive again.

— The banning of cows from the city of Delhi in India was the result of decision-makers not understanding the important role that these animals play in reducing municipal waste by eating most of the vegetable household residues, converting waste to dung for use as a fuel and contributing substantially to their owners' income by producing milk. Banning cows (rather than changing the way in which they are managed) causes problems in the system.

— Although from a technical point of view the optimal slaughter weight of hybrid pigs is 120 kg, a weight of 70 kg or less might be better-suited to the prevalent economic and environmental conditions. Pursuance of production objectives that originate in one country can be counterproductive and accelerate resource depletion in another country (or city/neighbourhood).

— Destroying cattle suspected of bovine spongiform encephalopathy (BSE) may serve political or economic ("consumer concern") purposes, but the real causes of the disease lie in the industrial way in which animals are kept (including the almost unrestricted movement of pathogens in resources such as bonemeal for animal feed).

— The dipping of animals and the use of hermetically closed buildings may prevent salmonella infections in poultry or pigs, but such measures also decrease the animals' resistance and are costly. Simpler housing and hygienic measures that allow for a certain rate of infection demand lower investments and make the animals less susceptible to disease, as well as helping more people to stay in business. The salmonella risk can be overcome by teaching vulnerable groups how to cook their food safely.

— Digging wells to increase water availability instead of decreasing water requirements by adjusting herd size often leads in the long term to a complete, irreversible depletion of water resources.

— Legislation that promotes large-scale enterprises by providing subsidies can be less productive than giving support to community-based activities that can sustain themselves in the long term.

Indigenous and Exogenous Technology

The distinction between indigenous and exogenous technology is again not clearly defined, but there are significant differences between the solutions that are found by farmers themselves and those that are introduced from elsewhere. Some solutions from outside are very useful, such as vaccines against disease, new ways of conserving feed or new designs for biogas pits. Other exogenous technologies, however, are not adapted to the local conditions and it may be cheaper and more practical to use locally developed solutions. For instance, there is much that official veterinarians can learn from local medical wisdom, methods of feeding the animals, etc. The use of indigenous veterinary medicine does not provide a solution to all problems, but it is cheap and accessible, particularly for the resource-poor farmer.

"Scientific" chemical analysis in feed laboratories has sometimes helped to increase the efficiency of livestock production, but has also created communication difficulties between the research and extension community and farmers who have other measures of feed quality such as cost, palatability and dustiness. Farmers judge feed quality, for example, on "leafiness" and toughness of the stem, rather than by referring to levels of crude protein or crude fibre. Proper communication and participation helps identify common approaches in seemingly divergent traditions.

On-The-Shelf Technologies and Management Options

Urban livestock systems are microcosms of systems that occur elsewhere. For example, goats, intensive poultry and high productivity dairy cows are commonly kept in both urban and rural settings in several places in the world. "On-the-shelf" technologies are therefore reflected in farmers' practices, fields and minds as well as in research stations. The following examples of such technologies are ready for niche application in urban conditions:

- breeding methods;
- housing systems;
- matching of feed resources for more efficient production;
- heat detection methods;
- veterinary methods for diagnosis and curative action;
- small- and/or large-scale processing of food and waste.

Many of these technologies, however, have only limited application unless they are suited to local economic and social conditions. In general, on-the-shelf technologies need to be refined through local participation rather than with laboratory work that does not represent real-life conditions and that may not even be used to report problems during the trial or adaptation period.

References

Arias, E., "Linking students to urban livestock producers in Mexico City", *Urban Agricultural Magazine*, 1(2). Leusden, the Netherlands, RUAF, 2000.

De Haan, C., Steinfeld, H. & Blackburn, H., *Livestock and the environment: finding a balance,* Fressingfield, Eye, Suffolk, UK, WREN Media, United Nations, United States Agency for International Development and the World Bank, 1996.

FAO, "Livestock's contribution to the protection of the environment", by C. Dalibard, *World Animal Review*, 1995.

______________, *Preliminary observations on peri-urban livestock production systems, Damascus, Syria,* by J. Mäki-Hokkonen, Mission report, FAO-AGAP, Rome, 1996.

Habitat, *An urbanising world: global report on human settlements*, United Nations Centre for Human Settlements, HABITAT, New York, Oxford University Press, 1996.

Smith, O.B. & Olaluku E.A., "Peri-urban livestock production systems", *Cities Feeding People Report Series,* 24. Québec, Canada, IDRC, 1998.

6

Biotechnology in Livestock Production

Foods derived from GM livestock and poultry are far from commercial use. Research efforts are underway at a large number of laboratories across the world to enhance the quality of live stock for human consumption. While it may take some time that the actual commercial application of these efforts really start and has a significant impact on human development, efforts are in the right direction and may lead to path breaking results sooner or later.

Research is also underway to enhance the quality of poultry such as chicken and an early detection of virus as well as developing immunity to virus in poultry. This has both human and commercial angle as virus free chicken is fit for human consumption and will address the misgivings relating to SARS virus outbreak in a large number of countries. It is also expected to enhance the poultry farming and lesser mortality in poultry leading to better returns for poultry farmers.

Biotechnology in the Broiler Industry

The poultry industry operates in a complicated and complex environment. There are many external influencing factors that, as responsible breeding companies, we must listen to and respond to where necessary. Breeding programmes must have a sound ethical basis, they need to be sustainable and they need to be commercially competitive. Whilst efficient poultry meat production is vital, it is also the industry's responsibility to ensure that this end is achieved through the highest poultry health and welfare standards as well as rigorous attention to food safety.

Promoting high standards of health and welfare will deliver significant benefits to the broiler grower. For example, there is less need to administer antibiotics therefore flocks will perform more predictably and production will be more efficient. Human food safety requires rigorous flock monitoring and testing to ensure the absence of disease, effectiveness of vaccinations and tight biosecurity.

Biotechnology has made significant contributions in these areas and continues to deliver real economic benefits for broiler growers. Advances in microbial biotechnology have led to the manufacture of more effective vaccines, enzymes and dietary ingredients. For example, recombinant DNA technology is being used to produce novel vaccines that are multivalent—so are effective against multiple pathogens or several strains of the same pathogen. Multivalent vaccines should have less risk of reversion and they are commonly accepted as being more effective than natural variants.

The same technology delivers key benefits through the manufacture of a new and wider range of enzymes such as Xylanases or Proteases. These enzymes have improved poultry gut function allowing the broiler to extract more nutrients, including nitrogen and phosphorus, from feed thus opening up the possibility of a more cost effective diet without having to compromise on quality. The improved availability of minerals has benefits for the welfare of broilers, for example it allows better skeletal development, as well as minimising the impact broiler production can have on the environment. Improvements in controlling heat stability of the enzymes mean that these enzymes can be delivered in the broilers' feed—a non-invasive application process.

These improved enzymes are so effective at producing nutrients that research is already underway developing new future applications—for example litter management.

Plant Biotechnology

Plant biotechnology is well understood and has a long history using conventional technologies, however more recently transgenic or GM technology has been used to introduce herbicide tolerance or resistance to pests or diseases. Plant biotechnology may in the future lead to advances in the development of vaccine varieties more appropriate as animal feed raw materials. For example, plant varieties may be developed with improved vitamin or mineral content or an amino acid balance closer to the requirements of growing broilers. Whilst the principle of genetically modified products remains controversial, advances in plant biotechnology are continuing to be made. If investment in developing new GM crops continues to exceed investment in developing new conventionally bred varieties, the differential in commercial performance will continue to widen.

Looking specifically at poultry biotechnology there are a number of techniques that biotechnology can offer. Some are well established and others have yet to be accepted.

In Ovo

In ovo (in the egg) biotechnology is advanced, it is widely accepted and practiced. The technology is easily managed and its very nature makes it precise and efficient—this technology has already replaced the conventional administration of vaccines by injection of newly hatched broiler chicks. In ovo administration provides a precise dose of vaccine resulting in early protection to disease challenges in the field, significant welfare benefits for the chicks and management benefits for growers. In addition, vaccine compatibility continues to improve enabling multiple applications in a single dose—making vaccination administration less invasive and resulting in benefits in terms of cost effectiveness and animal welfare. Already more than 5 billion broilers per year receive in ovo vaccination and the future trend is upwards. Development continues to improve the vaccine product range available to be administered in ovo.

Embrex Inc. is developing a new technology to determine the sex of an embryo before hatch. At Aviagen we are anticipating the development of gender-sorting using in ovo technology within the next two years. This technology will enable us to provide our customers with gender-sorted flocks that are handled significantly less and as a result will receive less stress in the vital early days of life.

Future advances in this technology will bring further benefits for growers while improving the health and welfare of the chicken.

Genomics

Genomics is the study of the DNA sequence, how it organised and how it functions. There have been major advances in sequencing the chicken genome and we expect a full sequence to be published in 2004. In the meantime more than 2,500 DNA markers have been identified within the poultry genome map—enough to detect major genes affecting traits of interest (Quantitative Trail Loci (QTL)) that make Marker Assisted Selection (MAS) a reality. For example, Chromosome 2 contains regions with large effects on growth, fat content and white meat yield. By selecting for a combination of physical measurements and the presence of particular DNA markers we can be more accurate and efficient. Advances in this technology will facilitate development of our products, to best meet changing needs of customers and markets worldwide.

However, the discovery of QTL does not identify the genes involved and it is only of limited value for changing those traits that are difficult to measure. Candidate gene MAS supports research into specific genes of interest and is significantly more efficient in identifying traits that are hard to measure or have low heritability, for example disease resistance or immune system function. Continued research in this area will reap significant genetic gain and provide a product that is more robust and consistent in its performance.

Delivering Sustainable Products

Developments in another widely accepted area of

biotechnology—functional genomics—go further, allowing the identification of the specific genes affecting traits of interest. With such identification come a number of benefits, for example, the selection for certain traits or product development to suit specific local environments. Whilst there are future opportunities in the area of association studies for developing candidate gene MAS techniques the challenge remains to ensure that the gene map relates to gene function. Advances in biotechnology will deliver the poultry genome sequence within the next couple of years.

It is important to take a balanced approach to selection—selecting for improved commercial performance while continuously improving the welfare of our birds will deliver a more long-term, sustainable product. Continuing advances in biotechnology make product development more accurate across a growing number of traits; general fitness, skeletal quality, cardiovascular function and disease resistance resulting in a more robust bird that can perform predictably in a wide range of environments—a product that has improved growth rate, feed efficiency, yield and robustness.

There continues to be debate surrounding the balance between improving welfare and improving commercial performance—biotechnological advancements are showing that it is possible to achieve both goals simultaneously. The British Poultry Council gait score survey (1994-2000) showed improvements in live weight, FCR, quality, livability and improved efficiency of poultry meat production. Similar data are available from the US industry.

No Transgenesis and Cloning

In the area of poultry biotechnology two other techniques—transgenesis and cloning—have been slower to develop and currently, the benefits remain less clear. Unlike the application of genomics through MAS, these techniques require invasive measures. Access to single celled embryos is only possible through surgery and the male and female pronuclei are not easily visible or accessible. This makes genetic manipulation and cloning in poultry more technically challenging than in farm mammals. Recent progress in establishing embryonic stem cells in poultry will make both technologies more practical.

Whilst Aviagen monitors the development of transgenic and cloning technologies, the company believes that real benefits for commercial broiler growers remains some way off. Currently, cloning is only possible on a small-scale experimental basis. At Aviagen, biotechnology research does not use transgenesis for a number of reasons. Ethical issues surrounding the technique remain unresolved and, to date, the company believes that the technology is limited with no genes of desired effect being available. Most importantly our customers are not convinced that transgenic or cloning will deliver real benefits.

In a highly competitive market, continued investment in biotechnology and research and development generally must remain a priority. There is a widening range of acceptable technologies, notably microbial, plant and poultry technologies, which already deliver benefits to the flocks and to the broiler grower. And there is more in the pipeline. Advances continue to lower the cost of production and improve the utilisation of raw materials that, in turn, reduces the impact on the environment. In short, the field of biotechnology continues to make significant contributions to bird health and welfare and the success of broiler production.

References

Sim, J.S. and S. Nakai, editors, *Egg Uses and Processing Technologies - New Developments*, CAB International, UK 1994.

Sim, J.S., S. Nakai, and W. Guenter, editors, *Egg Nutrition and Biotechnology*, CABI Publishing, UK 2000.

Cunnane, S.C. and L. U. Thompson, editors—*Flaxseed in Human Nutrition*, AOCS Press, Champaign, 1995.

Naidu, A.S., editor, *Natural Food Antimicrobial Systems*, 2000.

Sonaiya, E.B., "Poultry husbandry in small rural farms", *Entwicklung und landlicher Raum*, 1990.

7

Family Poultry Production Systems

Family poultry is defined as small-scale poultry keeping by households using family labour and, wherever possible, locally available feed resources. The poultry may range freely in the household compound and find much of their own food, getting supplementary amounts from the householder. Participants at a 1989 workshop in Ile-Ife, Nigeria, defined rural poultry as a flock of less than 100 birds, of unimproved or improved breed, raised in either extensive or intensive farming systems. Labour is not salaried, but drawn from the family household.

Family poultry was additionally clarified as "small flocks managed by individual farm families in order to obtain food security, income and gainful employment for women and children". Family poultry is quite distinct from medium to large-scale commercial poultry farming. Family poultry is rarely the sole means of livelihood for the family but is one of a number of integrated and complementary farming activities contributing to the overall well-being of the household.

Poultry provide a major income-generating activity from the sale of birds and eggs. Occasional consumption provides a valuable source of protein in the diet. Poultry also play an important socio-cultural role in many societies. Poultry keeping uses family labour, and women (who often own as well as look after the family flock) are major beneficiaries. Women often have an important role in the development of family poultry production as extension workers and in vaccination programmes.

For smallholder farmers in developing countries (especially in low income, food-deficient countries [LIFDC]), family poultry represents one of the few opportunities for saving, investment and security against risk. In some of these countries, family poultry accounts for approximately 90 percent of the total poultry production. In Bangladesh for example, family poultry represents more than 80 percent of the total poultry production, and 90 percent of the 18 million rural households keep poultry.

Landless families in Bangladesh form 20 percent of the population and they keep between five and seven chickens per household. In LIFDC countries, family poultry-produced meat and eggs are estimated to contribute 20 to 30 percent of the total animal protein supply, taking second place to milk products (38 percent), which are mostly imported. Similarly, in Nigeria, family poultry represents approximately 94 percent of total poultry keeping, and accounts for nearly four percent of the total estimated value of the livestock resources in the country.

Family poultry represents 83 percent of the estimated 82 million adult chickens in Nigeria. In Ethiopia, rural poultry accounts for 99 percent of the national total production of poultry meat and eggs. Poultry are the smallest livestock investment a village household can make. Yet the poverty-stricken farmer needs credit assistance even to manage this first investment step on the ladder out of poverty. Poultry keeping is traditionally the role of women in many developing countries.

Female-headed households represent 20 to 30 percent of all rural households in Bangladesh, and women are more disadvantaged in terms of options for income generation. In sub-Saharan Africa, 85 percent of all households keep poultry, with women owning 70 percent of the poultry.

Income generation is the primary goal of family poultry keeping. Eggs can provide a regular, albeit small, income while the sale of live birds provides a more flexible source of cash as required. For example, in the Dominican Republic, family poultry contributes 13 percent of the income from animal production. The importance of poultry to rural households is illustrated by the example below from the United Republic of Tanzania. Assuming an indigenous hen lays 30 eggs per year, of which 50 percent are consumed and the remainder have a hatchability of 80 percent, then each hen will produce 12 chicks per year. Assuming six survive to maturity (with 50 percent mortality), and assuming that three pullets and three are cockerels, the output from one hen projected over five years would total 120 kg of meat and 195 (6.8 kg) eggs.

Table 1. Annual budget for a family farm with 0.4 ha irrigated paddy, 0.1 ha vegetable garden, 100 ducks and two buffaloes in Indonesia

Annual expenses	Unit		Rupees
Crops			1,198,000
Animals:			
- Buffaloes			
- Ducks			
	Subtotal		1,147,200
			2 345 200
Annual revenue			
Crops:			
- Maize		240 kg	96,000
- Rice		4,000 kg	2 000,000
- Cassava		600 kg	60,000
- Peanut		60 kg	60,000
- Soybean		60 kg	30,000
- Mixed garden			
			150,000
	Subtotal Crops		2,396,000
Animals:			
- Buffaloes	- meat	150 kg	300,000
	- draft	30 days	180,000
	Subtotal Buffaloes		480,000
- Ducks	- eggs	13,140 eggs	5,256,000
	Subtotal Animals		5,736,000
Annual net return to family labour from crops			1,198,000
			(20.7%)

Annual net return to family labour from livestock	
- Buffaloes	480,000 (8.3%)
- Ducks	4,108,800 (71.0%)
Total return to family labour from agriculture	5,786,800 (100%)

A study on income generation in transmigrant farming systems in East Kalimantan, Indonesia (Table 1), showed that family poultry accounted for about 53 percent of the total income, and was used for food, school fees and unexpected expenses such as medicines.

Flock composition is heavily biased towards chickens in Africa and South Asia, with more ducks in East Asia and South America. Flock size ranges from 5 - 100 in Africa, 10 - 30 in South America and 5 - 20 in Asia. Flock size is related to the poultry farming objectives of:

— home consumption only;

— home consumption and cultural reasons;

— income and home consumption; and

— income only.

In Bangladesh, the average production rate per local hen of 50 eggs/year was regarded by some as low productivity. However, if it is considered that 50 eggs per hen per year represents four hatches from four clutches of eggs laid, incubated and hatched by the mother hen, and the outcome is 30 saleable chicken reared per year (assuming no eggs sold or eaten, 80 percent hatchability and 25 percent rearing mortality), then it is a remarkably high productivity.

Production Systems

Family poultry are kept under a wide range of conditions, which can be classified into one of four broad production systems:

— free-range extensive;

— backyard extensive;

— semi-intensive; and
— intensive.

Free-Range Extensive Systems

In Africa, Asia and Latin America, 80 percent of farmers keep poultry in the first two extensive systems. Under free-range conditions, the birds are not confined and can scavenge for food over a wide area. Rudimentary shelters may be provided, and these may or may not be used. The birds may roost outside, usually in trees, and nest in the bush. The flock contains birds of different species and varying ages.

Backyard Extensive Systems

Poultry are housed at night but allowed free-range during the day. They are usually fed a handful of grain in the morning and evening to supplement scavenging.

Semi-Intensive Systems

These are a combination of the extensive and intensive systems where birds are confined to a certain area with access to shelter. They are commonly found in urban and peri-urban as well as rural situations. In the "run" system, the birds are confined in an enclosed area outside during the day and housed at night. Feed and water are available in the house to avoid wastage by rain, wind and wild animals.

In the European system of free-range poultry keeping, there are two other types of housing. The first of these is the "ark" system, where the poultry are confined overnight (for security against predators) in a building mounted on two rails or skids (usually wooden), which enable it to be moved from place to place with draught power. A typical size is 2 × 2.5 m to hold about 40 birds.

The second type of housing is the "fold" unit, with a space allowance (stock density) for adult birds of typically 3 to 4 birds per square metre (birds/m^2), both inside and (at least this) outside. The fold unit is usually small enough to be moved by one person. Neither of these two systems is commonly found in developing countries.

Intensive Systems

These systems are used by medium to large-scale commercial enterprises, and are also used at the household level. Birds are fully confined either in houses or cages. Capital outlay is higher and the birds are totally dependent on their owners for all their requirements; production however is higher. There are three types of intensive systems:

— *Deep litter system*: birds are fully confined (with floor space allowance of 3 to 4 birds/m2 within a house, but can move around freely. The floor is covered with a deep litter (a 5 to 10 cm deep layer) of grain husks (maize or rice), straw, wood shavings or a similarly absorbent (but non-toxic) material. The fully enclosed system protects the birds from thieves and predators and is suitable for specially selected commercial breeds of egg or meat-producing poultry (layers, breeder flocks and broilers).

— *Slatted floor system*: wire or wooden slatted floors are used instead of deep litter, which allow stocking rates to be increased to five birds/m2 of floor space. Birds have reduced contact with faeces and are allowed some freedom of movement.

— *Battery cage system*: this is usually used for laying birds, which are kept throughout their productive life in cages. There is a high initial capital investment, and the system is mostly confined to large-scale commercial egg layer operations.

Intensive systems of rearing indigenous chickens commercially is uncommon, a notable rare exception being in Malaysia, where the industry developed in response to the heavy demand for indigenous chickens in urban areas. However, this accounts for only two in every 100,000 (0.002 percent) of that country's indigenous chicken.

Types of Poultry Species and Breeds

All species of poultry are used by rural smallholders throughout the world. The most important species in the tropics are: chickens, guinea fowl, ducks (including Muscovy ducks), pigeons, turkeys and geese. Local strains are used, but most species are not indigenous. The guinea fowl (Numididae) originated in West Africa; the Muscovy duck (Cairina moschata) in South America; pigeons (Columba livea) in Europe; turkeys (Meleagrididae) in Latin America; pheasants (Phasianidae) in Asia; the common duck (Anas) in Europe; and geese (Anser) in Asia.

Flock composition is determined by the objectives of the poultry enterprise. In Nigeria for example, the preference is for the smooth-feathered, multicoloured native chickens or Muscovy ducks. Multicoloured feathers serve as camouflage for scavenging birds against predators, including birds of prey, which can more easily see solid colours (especially white). Foundation stock is usually obtained from the market as grower pullets and young cockerels. A hen to cock ratio of about 5:1 is common. Both sexes are retained for 150 to 300 days, for the purposes of culling, selling, home consumption and gifts, most of which require adult birds.

In the last 50 years, there has been a great advance in the development of hybrid breeds for intensive commercial poultry production. This trend is most noticeable in chickens, turkeys and ducks. The new hybrids (those of chickens in particular) are widely distributed and are present in every country in the tropics, even in the most remote villages. The hybrids have been carefully selected and specialised solely for the production of either meat or eggs. These end-product-specialised hybrid strains are unsuitable for breeding purposes, especially for mixing with local village scavenger stock, as they have very low mothering ability and broodiness.

For the smallholder, keeping hybrids means considerable changes are required in management. These changes are expensive for the following reasons:

- All replacement day-old chicks must be purchased.
- Hatchery chicks require artificial brooding and special starting feed.

— Hybrids require higher quality balanced feed for optimum meat and egg production.
— Hybrids require more careful veterinary hygiene and disease management.
— Egg-laying hybrid hens require supplementary artificial light (a steadily increasing day-length up to 17 hours of total light per day) for optimum (profitable) egg production.

The meat and eggs from intensively raised hybrid stock are considered by many traditional consumers to have less flavour, and the meat to have too soft a texture. Consumers will thus often pay a higher price for village-produced poultry meat and eggs. Thus for rural family poultry keepers, it is more appropriate to maintain and improve local birds to meet this demand.

Chickens

Chickens originated in Southeast Asia and were introduced to the rest of the world by sailors and traders. Nowadays, indigenous village chickens are the result of centuries of cross-breeding with exotic breeds and random breeding within the flock. As a result, it is not possible to standardise the characteristics and productive performance of indigenous chickens.

There is no comprehensive list of the breeds and varieties of chickens used by rural smallholders, but there is considerable information on some indigenous populations from various regions. Most of this is based on feather colour and other easily measured body features (genetic traits), but more detailed data are becoming available. Examples of local chickens from different parts of the tropics are given in Tables 2 to 4 below. These evaluations were usually carried out under intensive management conditions in research stations, as the objective was to evaluate the local birds' productivity. More recently, data on the performance of local chickens under extensive management have become available, which makes it possible to compare performance under extensive and intensive systems (Table 4).

Table 2. Performance of local breeds in South Asia (intensively housed)

Traits	Desi	Naked Neck	Aseel	Kadak-anath	Black Bengal
12 wk live wt (g)	544	629	640	NA	433
Age at 1st egg (d)	208	NA	219	NA	200
Eggs/hen/year	116	104	100	80	NA
Egg wt (g)	46	45	51	39	49
Fertility (%)	81	80	55	90	86
Hatchability (%)	55	61	45	61	68

Characteristics such as adult body weight and egg weight vary considerably among indigenous chicken populations, although reproductive traits, such as the number of laying seasons per year, the number of eggs per clutch and hatchability are more consistent. Desi hens in Bangladesh range from 190 to 200 days of age at first egg (an easy measure of age-at-sexual-maturity), and they lay 10 to 15 eggs per season in 3 to 4 clutches (3 to 4 times) per year, with a hatchability of 84 to 87 percent (percent of eggs set).

Table 3. Local chicken breeds of Ethiopia

Traits	Tukur	Melata	Kei	Gebsima	Netch
24 wk body wt (g)	960	1000	940	950	1180
Age at 1st egg (d)	173	204	166	230	217
Eggs/bird.yr	64	82	54	58	64
Egg wt (g)	44	49	45	44	47
Fertility (%)	56	60	57	53	56
Hatchability (%)	42	42	44	39	39

Indigenous village birds in Ethiopia attain sexual maturity at an average age of seven months (214 days). The hen lays about 36 eggs per year in three clutches of 12 to 13 eggs in about 16 days. If the hen incubates her eggs for three weeks and then rears the chicks for twelve weeks, then each reproductive cycle lasts for 17 weeks. Three cycles then make one year. These are very efficient, productive and essential traits for survival.

Guinea Fowl

Guinea fowl are native to West Africa but are now found in many parts of the tropics, and are kept in large numbers under intensive systems in France, Italy, the former Soviet Union and Hungary. In India, guinea fowl are raised in parts of the Punjab, Uttar Pradesh, Assam and Madhya Pradesh, usually in flocks of a few hundred birds. Guinea fowl are seasonal breeders, laying eggs only during the rainy season, under free-range conditions. They are very timid, roosting in trees at night, and although great walkers, they fly very little.

Guinea fowl thrive in both cool and hot conditions, and their potential to increase meat and particularly egg production in developing countries deserves better recognition. The first egg is normally laid at about 18 weeks of age, and unlike many indigenous birds (which produce a single clutch a year), guinea hens lay continuously until adverse weather sets in. In West Africa, laying is largely confined to the rainy season.

Guinea hens under free-range conditions can lay up to 60 eggs per season, while well-managed birds under intensive management can lay up to 200 eggs per year. The guinea hen "goes broody" (sits on eggs in the nest) after laying, but this can be overcome by removing most of the eggs. A clutch of 15 to 20 eggs is common, and the incubation period for guinea fowl is 27 days. Domesticated guinea fowl under extensive or semi-intensive management in Nigeria were reported to lay 60 to 100 eggs with a fertility rate of 40 to 60 percent.

Domesticated guinea fowl are of three principal varieties: Pearl, White and Lavender. The Pearl is by far the most common. It has purplish-grey feathers regularly dotted or "pearled" with white. The White guinea fowl has pure white feathers while the Lavender has light grey feathers dotted with white. The male and female guinea fowl differ so little in appearance (feather colour and body weight [1.4 to 1.6 kg]) that the inexperienced farmer may unknowingly keep all males or all females as "breeding" stock. Sex can be

Table 4. Performance of local chicken breeds under scavenging and intensive management systems

System	Country	Breed	BodyWt (g)	EggN°	EggWt(g)
Scavenging					
Africa	Burundi	Local	1,500	75	40
	Mali	Local	1,170	35	34
	United Rep. Tanzania	Local	1,200	70	41
Asia	Indonesia	Kampung	2,000	35	-
	Malaysia	Kampung	1,430	55	39
	Bangladesh	Local	1,140	40	37
	Thailand	Thai	1,400	40	48
	Thailand	Betong	1,900	18	45
	Thailand	Samae	2,300	70	-
Latin America	Dom. Rep.	Local	1,500	100	38
	Bolivia	Local	1,500	100	-
Intensive					
Africa	Egypt	Fayoumi	1,354	150	43
	Egypt	Dandarawi	-	140	45
	Egypt	Baladi	1,330	151	40
	Nigeria	Local	1,500	125	36
	United Rep. Tanzania	Local	1,652	109	46
	Uganda	Local	1,500	40	50
	Zambia	Local	1,500	35	52
Asia	Bangladesh	Desi	1,300	45	35
	India	Kadakanath	1,125	80	40
	Indonesia	Ayam Nunukan	2,000	150	48
	Indonesia	Ayam - Kampung	1,350	104	45

distinguished at eight weeks or more by a difference in their voice cry.

Domesticated guinea hens lay more eggs under intensive management. French Galor guinea hens can produce 170 eggs in a 36-week laying period. For example, from a setting of 155 eggs, a fertility rate of 88 percent and hatchability of 70 to 75 percent, it is possible to obtain 115 guinea keets (chicks) per hen. In deep litter or confined range conditions, a 24-week laying period can produce 50 to 75 guinea keets per hen.

Ducks

Ducks have several advantages over other poultry species, in particular their disease tolerance. They are hardy, excellent foragers and easy to herd, particularly in wetlands where they tend to flock together. In Asia, most duck production is closely associated with wetland rice farming, particularly in the humid and subtropics. An added advantage is that ducks normally lay most of their eggs within the three hours after sunrise (compared with five hours for chickens). This makes it possible for ducks to freely range in the rice fields by day, while being confined by night.

A disadvantage of ducks (relative to other poultry), when kept in confinement and fed balanced rations, is their high feed wastage, due to the shovel-shape of their bill. This makes their use of feed less efficient and thus their meat and eggs more expensive than those of chickens. Duck feathers and feather down can also make an important contribution to income.

Different breeds of ducks are usually grouped into three classes: meat or general purpose; egg production; and ornamental. Ornamental ducks are rarely found in the family poultry sector. Meat breeds include the Pekin, Muscovy, Rouen and Aylesbury. Egg breeds include the brown Tsaiya of Taiwan Province of China, the Patero Grade of the Philippines, the Indian Runner of Malaysia and the Khaki Campbell of England. All these laying breed ducks originate from the green-headed Mallard (Anas platyrhynchos platyrhynchos). The average egg production of the egg breeds is approximately 70 percent (hen.day basis). The Indian Runner, Khaki Campbell, Pekin and Muscovy are the most important breeds in rural poultry.

Indian Runner

This is a very active breed, native to Asia, and ideal for free-range. It is a very good egg layer and needs less water than most other breeds, requiring only a basin in which it can immerse its beak up to the nostrils. It is the most graceful and elegant of all ducks on land with its upright carriage and slim body. It stands at an angle of about 80° to the ground but when startled can be almost perpendicular.

Khaki Campbell

Originally bred in England, this breed is derived from three breeds: the wild Mallard, the Rouen and the Indian Runner. The female has an overall khaki colour, and the male has a bronze-green head. The female is best known for her prolific egg laying ability, with an average of 90 percent (on a hen/day basis) with an average 73 gram egg weight.

Pekin

Originally bred in China, this attractive meat breed is favoured by commercial producers throughout the world. It is large and meaty with an upright stance and a broad round head. It has white to lemon-yellow plumage and a yellow skin. It is hardy, a reasonable layer, and grows rapidly. Although timid, it is docile and easily confined by low fences. It is well suited to both large, specialised duck farms and smallholdings. Pekin ducks are the major meat duck breed in Thailand, Malaysia, Philippines, the Democratic People's Republic of Korea and China.

Muscovy

This is not genetically a duck or a goose, but is more similar to the goose (Anseridae). It eats grass, as do geese, and has a similarly long egg incubation period of 36 days (compared with that of ducks - 28 days). It is popular in areas where there is little wetland rice production, since it does not require swimming water. The female Muscovy is an excellent brooding mother. It is often used as a foster brooder-mother for other species such as ducks, chickens and guinea fowls. It is a poor layer, producing only 30 to 40 eggs per year under extensive management.

The male Muscovy can become very large (4.5 to 5.5 kg) while the female is smaller (2.3 to 2.8 kg). The feather colouring is usually a combination of black and white, ranging from mostly black to mostly white. The male has characteristic red fleshy outcrops around the eyes called caruncles. The Muscovy is the predominant waterfowl in Africa and Latin America, as it thrives well under free-range conditions. Numbers are increasing in parts of Asia where lean, red meat is popular. When mated with breeds of domestic ducks, they produce infertile hybrid offspring ("mule" ducks). These mule ducks are a major source of duck meat in Taiwan Province of China.

A three-way cross-system is used for white mule duck production. Firstly, Pekin drakes are crossed with white Tsaiya ducks to produce a cross-bred female line called the Kaiya duck. These are then crossed with large white Muscovy drakes, usually by artificial insemination.

The resulting progeny is a mule duck, which is sterile but grows rapidly. It has good carcass composition with more meat and less fat than the Pekin. These three-way crosses have the added advantages of the high egg production of the Tsaiya, the high growth rate of the Pekin and the good carcass quality and meat texture of the Muscovy. Their white feathers are more valuable as down than those of darker-feathered ducks.

In most tropical countries, there are local duck breeds that have been selected to suit local conditions. They may not perform as well as improved breeds, but they do have the ability to survive and produce well under local extensive and semi-intensive systems. Setioko described three Indonesian ducks: Tegal, Alabio and Bali. Improved genotypes have been introduced and have either been crossed with local ducks or remained reasonably pure. There was some concern about the ability of the improved genotypes to survive under traditional farming systems.

Trials conducted in the Mekong River Delta by The Bin found that hybrid ducks raised for meat in rice fields were more profitable than the local ducks, even though they consumed more feed and cost more to buy initially. However, when raised for egg production in rice fields and on canals, the hybrids did not perform as well as the local ducks.

Geese

Geese are less important in family poultry production, except in China, where mainly local breeds are kept, except for a few European breeds such as the Toulouse and White Roman, imported for cross-breeding purposes. The great variety in breed size of geese permits their use under various management conditions. At the less intensive levels of production preferred by most family producers, smaller-sized birds (weighing approximately 4 kg, such as the Lingxhian or Zie breeds in China) are easier to manage.

Geese are high in the broodiness trait, and have a consequent low egg production of 30 to 40 hatching eggs (in three to five laying cycles) per year. At the other extreme are breeds of high fertility (and egg number), which are smaller and are selected specifically for use in breeding flocks for their lack of broodiness. Breeds such as the Zie may lay 70 to 100 eggs annually. The importance of the wide gene pool variety in China is significant for the Asian region in particular and for the world in general.

Pigeons

Pigeons are scavengers (not fed any supplementary feed) in most countries, living on the roofs of houses and treated as "pets" that do not need to be fed. They appear to prefer homestead compounds to fields. In some countries, they are eaten only for ritual purposes. They normally lay two eggs in a clutch, and the young birds (squabs) hatch after 16 to 17 days. The growing squabs are fed by their mothers on crop milk, produced in the mother's crop (first stomach). This enables young squabs to grow very rapidly. They reach maturity in three to five months at a body weight of 200 to 300 g for males, and 150 g for females. Adult pigeons are monogamous for life.

Local pigeons are specific to different regions in the tropics. Africa has five breeds, within which Chad has three local breeds. Asia and the Pacific have five breeds, with local breeds found specific even to the Cook Islands. Latin America and the Caribbean islands have only one breed. Europe has six breeds, two of which come from Belgium.

Turkeys

These birds are native to Latin America. The breeds kept by rural producers in the tropics usually have black feathers, as distinct from the white-feathered breeds that are raised intensively. Where there are no geese and ostriches, they are the largest birds in the farming system. Body weight ranges from 7 to 8 kg in males and from 4 to 5 kg in hens. They have good meat conformation, produce about 90 eggs per year and have medium to good hatchability. They are more susceptible to disease than either chicken or ducks.

References

Acharya, R.M. & A. Kumar, "Performance of local birds in South Asia", *Indian poultry industry yearbook*, 7th Edition, 1984.

Adene, D.F., "An appraisal of the health management problems of rural poultry stock in Nigeria", In E.B. Sonaiya ed. *Rural Poultry in Africa-Proceedings International Workshop* held at Ile-Ife, Nigeria, 1990.

Ahmed, N., *Backyard Poultry feeding systems in Bangladesh*, Asian Livestock XIX, 1994.

Aini, I., "Indigenous chicken production in south-east Asia", *World Poultry Science Journal*, 1990.

Bussi, B., *Disease as a constraint on productivity in extensively managed poultry*, University of Edinburgh, Edinburgh, 1987.

Chabeuf, N., "Disease prevention in smallholder village poultry production in Africa", *Proceedings International Seminar on Smallholder Rural Poultry Production*, 9-13 October, 1990.

8

Poultry Feeding and Nutrition

A regular supply of low-cost feed, over and above maintenance requirements, is essential for improved productivity in the three farming systems used in family poultry production:

- free-range—poultry roost in trees at night;
- backyard—poultry are confined at night; and
- semi-intensive—poultry are enclosed during the day in a very limited scavenger resource base.

When feed resources are inadequate, a few birds in production are better than more birds just maintained, but without enough food for production.

Extensive Systems

Farmers attempt to balance stock numbers according to the scavenging feed resources available in the environment in each season. Under the free-range and backyard systems, feed supplies during the dry season are usually inadequate for any production above flock-maintenance level. When vegetation is dry and fibrous, the scavenging resources should be supplemented with sources of minerals, vitamins, protein and energy. Under most traditional village systems, a grain supplement of about 35 g per hen per day is given.

There have been various approaches to utilising a wider base of feed resources for the flock. One is the use of poultry species apart from chicken. Waterfowl, especially ducks, may be distributed throughout the wetland rural areas, where they can feed on such resources as snails and aquatic plants in ponds and lagoons. Another approach is the integration of poultry with the production of rice, vegetables, fish and other livestock. An example is the combination of chicken with cattle, as practised by the Fulani of Nigeria, where the chickens feed on the ticks on the cattle as well as on the maggots growing in the cattle dung. Chickens raised in the cattle kraal (compound) weighed an average of 500 g more than those in the same neighbourhood but outside the kraal.

Semi-Intensive System

Under the semi-intensive system, all the nutrients required by the birds must be provided in the feed, usually in the form of a balanced feed purchased from a feed mill. As these are often expensive and difficult to obtain, smallholders use either unconventional feedstuffs or "dilute" the commercial feed by supplementing it with grain by-products (which supply energy and some protein). A well-balanced feed however is difficult to achieve, as grains and plant protein sources (the by-products of a few oil seeds) are becoming increasingly unavailable for livestock, and premixed trace minerals and vitamins are usually too expensive for smallholders.

Phosphorus and calcium can be obtained from ashed (burnt and crushed) bones; and calcium from snail shells, fresh or seawater shellfish shells, or limestone deposits. Salt to supply sodium can come from evaporated seawater or land-based rock salt deposits. These mineral sources are rarely used. Feed provided for birds kept under this system is therefore of a much poorer quality (unbalanced by dilution with crop by-products) than under either the extensive or fully intensive system.

The size and productivity of the village flock ultimately depend on the human population and its household waste and crop residues, and on the availability of other scavengable feed resources. There is a clear relationship between egg production and nutrient intake. This is demonstrated in Bangladesh, where fewer eggs are laid in the rainy season from August to September, but when snails are available in January and February, production increases.

The Scavengable Feed Resources Base (SFRB) include:

— household cooking waste;

— cereal and cereal by-products;

— roots and tubers; oilseeds;

— trees, shrubs (including Leucaena, Calliandra and Sasbenia) and fruits;

— animal proteins;

— aquatic plants (Lemna, Azolla and Ipomoea aquatica); and,

— commercially prepared feed.

The SFRB was defined as the total amount of food products available to all scavenging animals in a given area. It depends on the number of households, the types of food crops grown and their crop cultivating and crop processing methods, as well as on the climatic conditions that determine the rate of decomposition of the food products.

Seasonal fluctuations in the SFRB occur due to periods of fallow or flooding, cultivation, harvesting and processing. The SFRB includes termites, snails, worms, insects, grain from sowing, harvesting by-products, seeds, grass, fodder tree leaves, water-plants and non-traditional feed materials. The SFRB can only be harvested by scavenging animals, of which poultry are the most versatile, although this varies with species. Several types of poultry scavenging together can make more effective use of this resource.

Keeping poultry under the free-range and backyard systems depends to a large degree on the quality of the feed available from scavenging. Therefore it is essential to know what feed resources are available. For example: a flock of 12 young growing chickens with five productive hens have access to an SFRB of 450 g (dry weight) containing nine percent protein and 2,300 kcal of metabolisable energy (ME)/kg. This supports about 22 percent daily egg production, with about three eggs/clutch, assuming 80 percent of the SFRB was utilised.

Estimating Methods of SFRB

The value of the SFRB can be estimated by weighing the amount of daily food product/household waste generated by each family as parameter "H", which is then divided by the proportion of food product/household waste found in the crop of the scavenging bird (assessed visually) as parameter "p". This is then multiplied by the percentage of households that keep chickens (parameter "c"):

$$SFRB = H/p©$$

For example, an SFRB measured using the above method in Southeast Asia ranged from 300 to 600 g on a Dry Matter (DM) basis, containing eight to ten percent of vegetable protein and 8.8 to 10.4 megajoules (MJ) of metabolisable energy (ME) per kg (2,100-2 500 kilocalories [kcal] ME per kg). The amount of protein and ME in the SFRB was determined by analysis of the crop content. In Sri Lanka, the annual SFRB available to each family was calculated to contain 23 kg of Crude Protein (CP) and 1959 MJ of ME (468 mega [M] cal of ME). The crop contents are shown below after Table 1.

Table 1. Calculated values of SFRB for family flocks in different countries of Southeast Asia

Country	SFRB as kg DM/year
Indonesia	475
Thailand	390
Sri Lanka	195
Sri Lanka	197

The crop contents comprised:

— 72 percent household waste;

— 13 percent grass;

— 8 percent animal matter (earthworms, snails, ants and flies); and

—7 percent paddy rice.

For composition details of crop contents and food/products household waste, see Table 2 in the next page.

Table 2. Average composition of major feed components and crop content of scavenging hens in Sri Lanka

Component	DM	CP	EE	CF	Ash	Ca	P
			Percent			mg/g	
Food product/ household waste	43.2	10.3	7.2	2.2	1.4	0.8	4.0
Coconut residue	24.1	6.9	38.1	8.9	1.1	1.1	6.0
Broken rice	89.9	9.0	1.3	1.5	3.2	0.5	1.4
Crop content	34.4	9.4	9.2	5.4	16.0	0.8	0.9

Each family flock had access to the food product/household waste from two households, so that on average the amount available to the household flock was 550 g of Dry Matter per day. Daily egg production ranged from 11 to 57 percent, with an average of 30 percent. This did not vary significantly over the 12 months of the study. Chicken body weight at 20 days ranged from 41 to 100 g, and at 70 days from 142 to 492 g. Mortality up to 70 days was 65 percent. Losses were attributed to predators, particularly dogs, cats, mongooses, crows and other birds of prey. More than 90 percent of the hen's day was spent scavenging over a radius of 110 to 175 m. Cattle and goat pens were favourite scavenging areas.

Factors affecting the SFRB

Among the factors determining the size of the SFRB are: climate; number of households; number and type of livestock owned; crops grown; and the religion of the household. This was clearly illustrated in a Sri Lankan study, where results showed that the total biomass of the scavenging population was proportional to the SFRB. If the available SFRB is exceeded, then production falls (birds die and hens lay fewer eggs). If there is a surplus SFRB (such as a good harvest or fewer birds due to disease or stock sale), then production increases (more chicks and growers survive and more eggs are laid). Hence the SFRB available in a community determines the production potential of the poultry. If the SFRB is known, other factors affecting production can be identified and the benefits of providing additional inputs assessed.

Table 3. Amount of household waste, calculated SFRB and average flock biomass

Location - Village name	Month	House waste	SFRB		Flock biomass
		DM (g)	DM (g)	CP (g)	CP (g)
Galgamuwa I	March	143	260	20	91
Galgamuwa I	Sept	267	834	78	75
Galgamuwa II	March	543	639	63	83
Galgamuwa II	Sept	549	603	49	36
Ibbagamuwa	June	414	575	56	57
Ibbagamuwa	August	307	365	43	48

The maximum productive size of the village flock depends on the SFRB. To keep the flock size in balance with the available SFRB, it is necessary to set fewer eggs for incubation, cull unproductive birds and sell stock as soon as they are saleable. Production capacity should also be adjusted to match the seasonal variations in the SFRB. For example, during harvest time, when the SFRB is increased, extra chicks and growers may be reared, but at the end of the dry season birds may need to be culled, sold or consumed.

Feed Ingredients

The on-line and CD-ROM versions of the FAO searchable database Feeds and Feeding provide a full resource on this topic for all types of livestock, including poultry.

Cereals and Cereal by-products

Examples of grains for supplementing scavenging poultry include millet, sorghum, maize, and rice in the form of whole and broken grains. Amounts supplied are inadequate when using the surveyed estimate of 35 g supplement grain/ bird.day. This and the tannin content of sorghum have led to a search for alternative grains and the evaluation of agro-industrial by-products.

Dehulled Rice Grain

This can be used with vegetable and animal protein supplements for all types of poultry. Rough or paddy rice, off-coloured rice and broken rice have been used up to 20 to 30 percent in poultry rations. Rice bran has a moderate quality protein of 10 to 14 percent, approximately 10.4 MJ of ME/kg (2500 kcal of ME/kg), and about 11 percent Crude Fibre (CF). It is rich in phosphorus and B vitamins. Because of its high oil content (14 to 18 percent) it easily goes rancid. For this reason it should make up no more than 25 percent of the ration. This also applies to rice polishings. Rice bran usually includes rice polishings, but is often adulterated with rice hulls/husks, which are very high in fibre and silicon, and have a low nutritive value. Nevertheless, rice bran is still an important feed resource.

Maize Starch Residue (MSR)

This is a by-product of the extraction of starch from fermented, wet-milled maize, which is used as a breakfast cereal in West Africa. It usually has more than 16 percent Crude Protein, although the amount varies according to the maize variety and processing method.

By-products from Local Breweries

Brewer's grain and yeast have become common ingredients for poultry rations, but the process of drying the wet by-product can be very expensive.

Legumes and Legume By-products

Non-traditional legumes, such as boiled jack bean (Canavalia ensiformis) and sword bean (Canavalia gladiata), have been shown to be acceptable to laying hens, although they should not form more than ten percent of the ration because the sword bean is of low nutritive value. Winged bean (Phosphocarpus tetragonolobus) contains approximately 40 percent Crude Protein and 14 percent oil, and its overall nutritive value is very similar to that of soybean and groundnut cake for broiler meat chicken. Winged bean leaf foliage is also acceptable to laying hens. Unless the plant is grown with stake supports, the yield is very low, which makes its cultivation on a large scale less economical. However it is suitable as a feed and fodder crop for smallholder poultry.

Soybean (Glycine max)

This crop is being grown increasingly for human consumption. If the cotyledons (fleshy beans) are used for human food, the testa (bean-seed coat) is given to poultry. Raw soybeans heat-treated by boiling for 30 minutes and then fed to scavenging birds in amounts of up to 35 percent of the ration resulted in satisfactory performance in broilers and laying hens. In pullets and layers fed raw soybeans with no heat treatment as 12 percent of the ration, there was a significant reduction in body weight at 20 weeks, as well as a delay of four days in the onset of sexual maturity. The heat treatment destroys a trypsin inhibitor, which, if left intact, prevents digestion of raw soybean.

Cowpea (Vigna unguiculata)

This legume crop is grown solely for human consumption in Africa. Its by-products, especially the testa (seed coat), are used as a feed for small ruminants and have also been fed to poultry. The testa represents about six percent of the weight of the whole cowpea, but is usually discarded (in West Africa) when the cotyledons are made into a puree for a locally popular fried cake. With its crude protein content of 17 percent, its apparent metabolisable energy (AME) value of 4.2 MJ of AME/kg (1005 kcal AME/kg) and its mineral profile, cowpea testa should be a good feed resource, but the presence of tannin (53 mg/g) and trypsin inhibitor (12.4 units/mg) limits its utilisation. Cowpea testa should not make up more than ten percent of the total feed of a poultry ration.

Roots and Tubers

Cassava (Manihot esculenta)

This is grown in large quantities in Africa, Asia and Latin

America, both for human consumption and as a livestock feed. Cassava and its by-products (in the form of leaves, small tubers, pulp, peels, chaff, gari [fermented grated tubers], gari sievings, whole fermented roots and ensiled cassava meal) are used. The dried chips are high in energy and fibre but low in protein. In regions where cassava is used for human food, the peels are the most useful part of the cassava plant for feeding livestock. Amounts of 20 to 45 percent cassava peel meal (CPM) have been fed to chickens, but its use is limited because of the high content of the poison hydrogen cyanide (HCN), as well as high Crude Fibre, low protein content and dust.

There is a considerable range of HCN levels in cassava, according to variety. When cassava completely replaces grains in a ration, there is a consequent reduction in egg weight and a change in egg yolk colour. Whether or not there are negative effects on egg fertility and hatchability is not known. Cassava meal gives good growth in meat chickens, although protein and other nutrients must be carefully balanced.

Molasses or sugar may be added to sweeten the bitterness of the cyanide and thus improve palatability. Oilseeds such as full fat soybean can compensate for the high fibre and low protein content and for the dustiness. To remove the cyanide, detoxification methods include ensiling, sun-drying, air-drying, roasting, boiling and soaking. For smallholders, the most practical method is sun-drying. Palm oil can also moderate the effects of cyanide on poultry. Some "sweet" varieties of cassava are used in human food preparation, and these are often fed to poultry, particularly ducks.

Sweet potato (Ipomoea batatas)

Dried sweet potato forming up to 35 percent of the ration has been fed successfully to broilers and layers. The tubers are boiled before use, which overcomes any problems with dust or fungal growth from storage.

Oilseeds

Oilseeds in full-oil or partly oil-extracted form are a source of both energy and protein for extensive and intensive poultry systems.

Cotton (Gossypium spp.)

Glanded cotton seed cake (CSC) is a high-demand supplement fed to ruminants, but if available it can be fed in amounts up to 25 percent in the diets of layers and broilers without adversely affecting egg production and growth. Poultry are tolerant of the gossypol found in CSC, but it can cause an olive discolouration of egg yolks, which consumers do not like. Addition of 0.25 percent ferrous sulphate should be added routinely to laying hen rations containing up to ten percent CSC.

Sesame (Sesamum indicum)

The feed consumption and conversion rates for birds fed various forms of raw unhulled sesame seeds were better than those for birds fed dehulled but whole sesame seeds, confirming the practice of smallholders who use whole sesame seeds as a supplement for scavenging poultry. Sesame seeds should used in amounts between 20 and 35 percent of the ration.

Groundnuts (Arachis hypogaea)

Groundnuts may be used in the oil-extracted cake form to make up 8 to 24 percent of the ration. Mouldy groundnuts may contain toxic substances, the most dangerous of which is aflatoxin.

Coconut (Cocos nucifera)

Coconut meal can be used to form 50 percent of the ration, especially when combined with a high-energy source such as cassava meal. It is low in lysine, isoleucine, leucine and methionine.

Sunflower (Helianthus Annuus)

Sunflower seeds can be fed whole, or the decorticated meal can be used to replace groundnut cake and soybean meal and up to two-thirds of fishmeal. It has the highest sulphur amino acid content of all the major oilseeds.

Oil Palm (Elaeis guineensis)

Most oil palms are processed locally. The by-products are kernels and an aqueous solution of oil, fibre and solids. This solution can be filtered to remove the fibre. This leaves an aqueous mixture called palm oil sludge (POS), which supplies feed energy and fatty acids. Sludge processed using chemical solvents should not be used, as the chemical residue may be toxic to the birds. It can be fermented and used in smallholder poultry systems or dried to form up to 40 percent of commercial compound feeds. Palm kernels are processed locally into palm kernel oil by heat or cold-water extraction.

Other oilseeds

Other oilseeds that have been fed to poultry under research conditions include rubber, amaranth, Niger seed (Nueg), breadfruit (Artocarpus altilis), locust bean (Ceratonia siliqua), African oil bean, melon, mango and castor oil. Okra seed (Hibiscus esculentus) has not yet been evaluated as a protein source for poultry, and although it is lower in protein, it compares favourably with soybean in all other nutrient components. Since okra is widely grown by smallholders and the seeds are kept for planting, it may be a potential source of protein for smallholder poultry.

Bambara groundnut (Voandzeia subterranea)

This is a good source of protein with a high lysine content. As the nut is not widely eaten, the plant is grown mainly as a mulch crop and the foliage is scavenged by poultry.

Trees, Shrubs and Fruits

Neem leaves

A pilot study was undertaken to test the response of three groups of layers to neem leaves. One group was fed a ration containing ten per cent fresh neem leaves, the second a ration of ten per cent dried neem leaves, and the third none. The group receiving the fresh neem leaves had increased feed intake, daily egg production and egg weight compared with the other two. There appears to be a fat component of fresh neem leaves (Azadirachta indica) that enhances egg production and egg weight.

Coffee pulp

This is high in fibre, but as the essential amino acid content is similar to that of soybean, it can only be used in limited amounts.

Citrus pulp

No more than two percent citrus pulp should be included in the ration to avoid reduction in growth rate and off-colour egg yolks.

Over-ripe bananas and plantains

These are of greater palatability for poultry than green bananas, which contain free or active tannins.

Derinded sugarcane pith and molasses

Sugarcane juice can make up to 25 percent of the poultry ration and molasses up to 30 percent, but it should be noted that over ten percent molasses results in watery faeces. Raw sugar however can be fed at up to 50 percent of the ration without watery faeces. Combining one part molasses with three parts sugar gives good production without the digestive problems. Molasses is often added to rations at low levels of inclusion to make it more palatable, although there may be problems with evenly mixing the liquid, and with fungal toxins in the stored feed, encouraged by the sugar levels.

Animal Protein

Blood meal

This is recognised as a high crude protein source with an imbalanced, relatively poor amino acid profile. Handling and processing of blood is difficult in low-technology situations. For processing small amounts, one method is to absorb the blood on a vegetable carrier such as citrus meal, brewers grain, palm kernel, ground maize, cob rice or wheat bran, after which the material is spread out for drying on trays heated from below or placed in the sun. At the farm level, the blood may be supplied from the slaughter of livestock. Abattoirs and slaughterhouses provide large volumes of blood for making up feeds at the commercial level.

Termites

Farina, et al., described a technique used to collect termites for scavenging poultry. Briefly, the straw of sorghum, millet and maize are chopped, placed in clay pots or calabashes and moistened. The mouth of the container is placed over a hole in a termite colony under construction. The container is covered with a jute sack to prevent drying out and a heavy stone is placed on it to secure it in position. After three to four weeks, a new colony of termites should be established inside the container. The eggs and larvae are particularly relished by chicks, guinea keets and ducklings, while adult birds also feed on the adult insects. Cattle dung can be used in place of the cereal straw.

Maggots

Alao and Sonaiya grew maggots on cowpea testa (seed coats) and monitored the chemical composition of the mixture over ten days. Cowpea testa samples were placed in a basket near a pit latrine to attract flies to lay eggs on them. Every two days, a sample was steeped in boiling water to kill the maggots. They were then sun-dried and ground before analysis. Results showed that the Crude Protein content of the mixture doubled by the second day.

Soukossi produced maggots from fibrous vegetable material and poultry droppings. The method was developed for feeding fish, but can easily be adapted for smallholder poultry. A tank with a capacity of one cubic metre is filled with water to about 15 cm from the top. Dried stalks of maize, amaranth, groundnut, soya and other legumes are soaked in the water to which some poultry droppings are added. Flies and other insects are attracted to the soaked material to lay their eggs. After five to seven days, eggs are hatched and larvae are sufficiently developed to be fed to fish. Beyond this period the maggots develop into adult flies. It was observed that up to 50 percent of the eggs laid by flies died if exposed to the sun for several hours. A cover, at least for the hottest hours of the day, is therefore necessary. Similar trials have been carried out in Burkina Faso.

Earthworms

Vorster et al. produced earthworms as a source of protein for chicken feed. In an area of 25 m^2, one kg of fresh earthworm biomass was produced daily. This is sufficient to supplement at least 50 chickens with high-quality protein. It must be noted, however, that earthworms (and snails as well) may be important vectors for tapeworms such as Davainea and Raillietina and also contain a growth inhibitor.

References

Leeson, S. & Summers, J.D., *Commercial poultry nutrition*, University Books, Guelph, Canada, 1991.

Willin, E.S., "Relation between egg weight and intensity of growth in geese", *Preliminary Proceedings, 10th European Symposium on waterfowl*, World's Poultry Science Assoacition, Halle, Germany, 1995.

9

Poultry Farm Management

Poultry farmers should always provide optimum condi-tions in the broiler house if they want to be among the topproducers and maximize on profits. The primary concerns of producers are house and bird temperature, ventilation,feed, water, litter conditions and overall bird health. All ofthese factors are critical to success in producing broilers andthey are topics which the company field service representa-tive will discuss with each grower during regular visits. Inaddition, there are several other areas around the farm thatwill need attention to maintain appearances and keep every-thing functioning properly.The first area that should get attention is weeds andgrass. First of all it is advantageous to have good grass coveraround farm building, particularly the broiler houses.

Grasshelps reduce reflected heat which in turn helps keep housetemperatures under control in hot weather. Well establishedgrasses also prevent soil erosion which in severe cases couldresult in damage to the house pad or the footing and founda-tion of the poultry house. Grass should always be kept mowedand weeds should also be kept under control. Not only areunmown weeds and grass unsightly, but they can encouragerodents and can also reduce the effectiveness of naturalventilation. Tall weeds and grass provide rats and mice aplace to hide that is close to the building.

Most rodent controlprogrammes will stress the need to keep areas around buildingsmowed frequently.Trees with low hanging limbs can also be problemsaround the broiler farm. The low limbs may not look too badfrom an appearance standpoint but they can certainly reduceair movement if they are too close to buildings. In very warmweather, when natural ventilation is a must, all low limbsshould be pruned and hauled away.

Housing and Runs

Under undomesticated conditions, poultry lay eggs in simple nests, perch in trees and spend much of the day scavenging for feed. Chickens spend a large proportion of their time scratching to expose hidden food. Under the backyard and semi-intensive production systems, poultry are usually enclosed at night to discourage thieves and predators, and under intensive production, are totally confined day and night. Some village households keep their few chickens inside the house or even under their bed at night, to discourage theft.

Given a choice of a place to lay their eggs, hens will choose a soft "litter" base, and they prefer an adequately sized (a cube of approximately 30 cm), darkened nest with some privacy. Prior to laying, hens usually investigate a number of possible sites before entering a nest box. They then show nesting behaviour, which includes a special protective nest-seeking voice, after which they sit and finally lay. When they have laid an egg, they announce this with another type of "pride of achievement" call. These calls can also be heard in a battery cage house. If perches are provided, hens will perch most of the time rather than stand on the wire floors, and after dark most birds roost on the perches. Perching is a probable survival characteristic to avoid night predators. The basic requirements for poultry housing are:

— space;

— ventilation;

— light; and

— protection (from weather and predators).

Space

This is the most important basic principle in housing, as the space available determines the number and type of poultry that can be kept. For example, a deep litter house measuring 6 m by 11 m can hold 200 laying hens at a stock density of 3 birds/m2 (3.6 ft^2/bird). Under the older system of measuring, stock density was measured in ft^2 per bird, which is the inverse of birds per m^2 used in the metric system, incorporating a conversion factor of 0.0929 m^2/ft^2.

Linear space or length of perch per bird is measured in centimetres. The recommended floor and perching space for the three main types of chicken is shown in Table 1.

Table 1. Requirement of chickens for floor and perch space

Chicken types	Floor Space (birds/m^2)	Floor Space (ft^2/bird)	Perch Space (per bird)
Layer	3	3.6	25 cm (10 in)
Dual Purpose	4	2.7	20 cm (8 in)
Meat	4-5	2.1-2.7	15-20 cm (6-8 in)

Hen groups are comfortable at a stock density of three to four birds per square metre. If more space is allowed, a greater variety of behaviour can be expressed. Less space creates stressed social behaviour, allowing disease vulnerability and cannibalism and leaving weaker birds deprived of feed or perch space. Individual birds need more room for normal behaviour and adequate exercise than the 22 birds/m^2 (0.5 ft^2/bird) density currently used in commercial laying cages. Over recent decades, animal welfare concerns have encouraged research on laying cage structures to make designs better suited to the needs of hens, while retaining cost-effectiveness for production.

Ventilation

Ventilation is an important factor in housing. A building with open sides is ideal, otherwise cross-ventilation at bird-level should be allowed for in the form of floor level inlets, open in a direction to allow the prevailing wind to blow across the width of the building. An air mass between the side walls of a poultry house resists being moved, even across an open-sided building. The wider the building, the more the resistant it is to air movement. Buildings over 8 m (26 ft) wide have a significantly greater problem because of this inherent property of air to resist movement. It is recommended that buildings relying on natural airflow for ventilation should not exceed 8 m in width.

Heat stress is a significant constraint to successful production and can lead to death. Although birds can withstand several degrees below freezing, they do not tolerate temperatures over 40°C. This depends on the relative humidity prevailing at the time. Poultry do not possess sweat glands and must cool themselves by panting out water in their breath, which is evaporative cooling. When the humidity is too high, this cooling mechanism does not work very well. Lethal temperatures for most chickens are 46°C upwards, and severe stress sets in above 40°C.

In temperate regions, the chicken house may be constructed to face the rising morning sun to gain heat. In the tropics however, an east-west orientation of the length of the building helps to minimize exposure to direct sunlight. Building materials such as tin or other metal should be avoided for this reason, although white paint will reflect up to 70 percent of incident solar heat radiation. Ventilation concerns in building alignment may prevail over solar heat control in this aspect, as cross-flow ventilation requires the side of the building to face the prevailing wind.

Ground cover can also reduce reflected heat. Shade should be provided, especially if there is little air movement or if humidity is high. With no shade, or when confined in higher temperatures, poultry become heat stressed and irritable, and may begin to peck at one another. When new pinfeathers are growing (especially on young stock), blood is easily drawn, which can lead to cannibalism. The effects of heat stress are:

— a progressive reduction in feed intake as ambient temperature rises;
— an increase in water consumption in an attempt to lower temperature;
— a progressive reduction in growth rate; and
— disturbances in reproduction (lower egg weight, smaller chicks, reduced sperm concentration and an increased level of abnormal sperm in cocks).

Light

A well-lit house is essential. A dark house leads to lethargic, inactive, unproductive birds. Light is important for feeding, as poultry identify food by sight. This is especially important for intensively managed day-old chicks, which need very bright 24-hour lighting for their first week of life.

Light is also an important factor in sexual maturity. An increasing light proportion in the day, as naturally occurs from mid-winter to mid-summer, will accelerate sexual maturity in growing pullets, bringing them to lay sooner. If hens are already laying, the increasing light proportion will increase egg production. The opposite effect is also true: as the light proportion of the day decreases (as naturally occurs from mid-summer to mid-winter), then sexual maturity is slowed in growing stock, and egg production is reduced in laying hens. These effects are somewhat reduced towards the equator, as the difference in the daylight proportion of a day changes less and less.

This physiological effect on poultry is important in terms of maintaining egg production in commercial flocks, and requires supplementary lighting programmes. Regular and reliable electricity supply is required for such programmes, otherwise the effect can be made worse by breaks in the light supplementation system. A slow but steady increase maximises the rate of production. However, lighting programmes producing an effective daylight proportion in excess of 17 hours per day can have a worsening effect on egg production. A 24-hour security lighting system can have such an effect on egg production.

Birds do best in situations where there is plenty of natural light that does not raise the temperature of the house. Natural light is preferable unless regular, reliable and well-distributed artificial light can be provided. It is recommended that the interior of the house be whitewashed to reflect light. The intensity or brightness of the light is also important. Egg production will decrease at light intensities lower than five lux (the "lux" is the metric unit of light intensity and can be measured by a meter similar to that measuring light intake into a camera lens), although meat chicken will keep growing optimally at light intensities as low as two lux (not bright enough to read a newspaper). These intensities are measured at the eye-level of the bird, not near the light source. Unless supplementary lighting is spaced uniformly, there may be areas in the building insufficiently lit to allow optimum growth or egg production. Designs for layout assume that the light bulbs or tubes will be kept clean, as dusty surfaces will reduce light output.

Protection

Many factors influence the type and choice of housing to protect poultry from the effects of weather and predators. These include the local climate, the available space, the size of the flock and the management system. In extensive systems, birds must be protected from disease and predators but also be able to forage. Traditional large animal fencing using live plants is not enough protection against predators such as snakes, kites, rats and other vermin.

A simple and effective system to deter predator birds is to tie parallel lines of string across the main scavenging area, the intervals between which measure less than the predator's wingspan; or, alternatively, a fishing net supported on poles can be spread across the side of the run where predator birds could swoop on the scavenging chicks.

Leg traps can be set for large predators. It is not necessary to set traps around all the pens, as predators tend to attack the same pen on the second night. Steel traps can be boiled in walnut hulls or cocoa pods, both to camouflage them and to prevent rust. The traps will be more effective if not touched with bare hands, as most predators have a keen sense of smell. Instead, they should be handled with a stick, rubber gloves or tongs.

Rats, mongooses and snakes are only a problem when the birds are small. Rats often come up through the earth floors, and the first signs of a rat attack may be unusually quiet chicks huddled under the brooder heater or in a corner, or dead chicks with small bloody neck scratches. Snakes will kill chicks if they can get into the brooder house. A treble fishhook in a dead bird can be left as bait: the snake will

Table 2. Predator attack modes and control methods

Predator	Attack mode	Control method
Hawk	Picks up stray birds and weaklings. Attacks birds so that head and toe marks are visible on back. Often plucks birds.	Hunt the hawk and keep chicks away from clear swoop areas.
Rat, mongoose	Usually take more than they eat, and stuff chicks in holes for later consumption.	If allowed, use rat poison.
Snake	Will swallow eggs and chicks	Use fishhooks.
Dog, cat	General destruction	Try to catch them. Cats can control rats but wild cats and dogs are a problem.
Fox, jackal	Will bite off the feathers over the back and between wings, eat the entrails and breast, and carry the bird to den.	Roam in the early morning; kill for their young. Trapping is best control.
Raccoon	Pulls off head and eats crop. Will carry birds off.	May be protected in some countries. A permit to destroy may be required.

swallow the hooks as it gulps down the bird and eventually die. Holes around doors and windows through which rats and snakes may enter should be plugged.

Coops or baskets may be used to house mother hens and chicks in order to reduce chick mortality due to predators, thieves and rain. They also allow for separate feed and water supplementation, although the inadequate feed usually provided in coops means that some scavenging remains necessary.

Housing in Free-range Systems

Overnight shelter which is roomy, clean and airy should be provided under free-range systems. Houses may be either fixed or mobile. If space permits, a mobile chicken house may be appropriate, and to increase egg production, mobile folds or field units for laying birds can be provided. These mobile units can be rotated on the range. Although housing is cheaper and there is less need for balanced rations, the birds are exposed to the sun and prone to parasite infestation.

The stocking density on pasture should be calculated according to the soil type and pasture management system. A night shelter for up to 20 free-range chickens can be attached to any existing structure, such as the farmer's outhouse, kitchen or dwelling. In a deep litter system, there should be a density of at most three to four birds per square metre. In regions where it rains heavily, the floor should be raised with a generous roof overhang, particularly over the entrance. The raised floor can be a solid platform of earth or a raised bamboo platform. The raised bamboo platform has the advantage of providing ventilation under the poultry, which helps cool them in hot weather and keeps them out of flood water in the monsoons.

The walls of the building can be made of mud or bamboo, and the windows and door of bamboo slats. The house can also be free-standing, and may also be suitable for semi-intensive or intensive production systems.

Housing in Semi-intensive and Intensive Systems

Planning

Complete confinement is only advisable where:

- there is good management;
- reproduction is spread equally over the year;
- land is scarce or inaccessible all year round;
- balanced rations are available;
- a supply of hybrid day-old chicks is available;
- labour is expensive;
- parasite and disease control are readily available; and
- the objective is commercial production.

The reasons for confinement are, in order of priority, to:

- reduce mortality due to predation in chicks under two months of age;
- achieve higher daily gain and better feed conversion in growers; and
- allow better supervision of production in laying hens.

In all confined systems, the location and building design must be carefully considered. The area surrounding the house should be mown or grazed. A good location should meet the following criteria:

- It should be easily accessible.
- There should be a reliable water supply.
- The ground should be well drained.
- It should be at a sufficient distance from residential areas (far enough to protect human health and close enough to provide security for the birds).
- It should be well away from woodland.

Converting existing facilities can provide housing, although planning permission may have to be obtained. An unused outhouse kitchen, for example, can be converted into a poultry house. In all conversions, maximum use should be made of the space available through careful planning:

— A plan of the building should be drawn to scale.

— Use should be made of existing floors and walls, if suitable.

— Space requirements of the birds and manure disposal should be taken into consideration.

— A feasibility study should be carried out, taking into consideration future plans and requirements as well as the economics of converting the building.

Construction

The floor is extremely important. An ideal floor for a deep litter house is well drained and made of concrete, with a layer of heavy gravel or wire mesh embedded in it to keep out rats. This type of floor is usually costly. Wood, bamboo, bricks or large flat stones (according to what is locally available) can be used, but are harder to clean. Clay floors are cheaper, but require the application of a fresh layer of clay either between flock batches or at least annually. In areas where construction materials are cheaper than deep litter, and particularly in humid regions where litter material is not available, raised floors are sometimes used.

These are made of wire mesh, expanded metal, wooden slats or split bamboo, to allow the droppings to collect under the house, and should be about one metre above the ground to allow for cleaning and ventilation. Higher floors may result in an unstable building. They are supported by pillars, which are either rot-resistant or have stone or concrete footings, and which are made of such materials as wood, bamboo, oil drums and concrete blocks. Houses with raised floors on posts can be protected against rats with baffles. The baffles can be made of a metal collar, a tin can turned upside-down or a metal band wound around the post, but must fit tightly to deter even the smallest rodent.

The roof and walls of the house can be made of any inexpensive local material, including bamboo slats, sorghum stalks, mud, wooden slats and palm fronds, as long as the structure is made relatively rat-proof. In colder regions, the walls should be thicker or insulated, but in warmer climates thatch can be used, although it should be replaced frequently to minimize parasite and disease problems. The inside of the walls should be as smooth as possible, to prevent tick and mite infestation and to make cleaning easier. Interior lengthways building partitions are not advisable, as they reduce cross-flow ventilation.

The roof should be watertight, and should overhang the walls by one metre if the windows have no shutters. The roof can be made of thatch, sheet metal or tiles. Thatch is usually the cheapest option and provides good insulation. It will probably have to be replaced every three years, or immediately if ticks get into it. It should be interlaced with bamboo or wooden slats to keep predators out. Sheet metal is usually too expensive, and in hot climates must be painted with white or aluminium to reflect sun heat. However, it is easily cleaned which is an important advantage where ticks are a problem. A layer of plastic sheeting sandwiched between bamboo slats is a good seal against rain and vermin. Flattened oil drums can be used at a lower cost. Although usually more expensive than thatch, sun- or oven-baked tiles will last much longer. Because of their weight, the frame for a tiled roof must be stronger than for other materials.

Window design depends on the local climate. Chickens need more ventilation than humans, but should be sheltered from wind, dust and rain. During storms, wood or bamboo hinged shutters or curtains made from feed sacks can cover window openings on the windward side of the house. In humid climates, window design should take as much advantage of the wind direction as possible to reduce the amount of moisture in the house. Window areas are best covered by wire mesh or expanded metal. Wooden slats or bamboo can be used, depending on available funds and materials. However, the thicker the material, the more ventilation will be reduced. Doors should be made of metal, wood or bamboo. The top half of the door could be wire mesh. Doors should be sufficiently strong to withstand being opened and closed many times a year.

Gabled roofs reduce solar heat loading when compared with flat or lean-to roofs. The pitch or "angle of rise" on a gabled roof is important for many reasons. Traditional village thatched gabled roofs are usually constructed using bush timber, and at a steeply pitched angle (greater than 42° from the horizontal), which helps the roof to withstand stormy winds. Shallower pitched roofs are more susceptible to being blown off in strong winds, particularly when the pitch angle is 15° to 20°. Shallower pitched roofs have less roof surface area, which reduces the cost of surfacing material, but because they are more affected by stormy winds, they need stronger support frames, which results in a much higher overall roof cost. A 42° pitch is the optimum compromise between roof surfacing costs and roof support costs.

The maximum width for an open-sided poultry building, under conditions of a slight breeze, which allows air movement across the shed at the height of the bird, is 8 m (26 ft). To maximise the volume and velocity of airflow across the shed width, the end walls of the shed should be closed. This forces the air to flow across the shed width, especially if the wind is not already coming from that side. Centre ridge ventilation is not recommended, as it discourages airflow across the full shed width. Air enters at the prevailing wind side and is drawn up at the centre to exit at the ridge, excluding the other half of the building.

Nests

To avoid excessive competition and minimize eggs laid on the floor, one nest should be provided for every five hens. If larger communal nests are used, at least one square metre per 50 birds should be allowed. Baskets, pots and cardboard boxes can be used for nests. Dimensions suitable for a basket or pot nest are a 25 cm base diameter, 18 cm high walls, and a 40 cm open top diameter. Nests should be situated in a secure, shady secluded place out of the sun, lined with fresh litter and kept clean. Nest boxes for individual hens should be constructed in multiple groups for larger numbers of hens. These are usually made of wood, and should measure approximately 30 cm on all sides, with a nest floor area of about 0.1 m^2.

Perches and Roosts

Chickens prefer to roost at night on perches. Perching space of 15 to 20 cm should be allowed for each bird. Birds lower in the social peck order can also use the perches during the day. The cross-section of each perch bar should be 2 to 3 cm. Their length depends upon the number of birds to be housed. The perches should be placed within a frame, and aligned parallel to the wall, and horizontally, with a sliding, removable platform called the "droppings board" about 20 cm below the perches to catch the manure droppings. The first perch bar should be placed 20 to 25 cm out from the back wall, and subsequent ones at 30 to 40 cm intervals.

The droppings board should touch the back wall and extend 30 cm in front of the front perch bar, as this will allow the birds to land from their flight from the floor before seeking a perching spot. Droppings boards should be a maximum of 75 cm from the floor of the house, and the perch bars should be about 20 cm above the droppings board, to facilitate cleaning of the droppings board. Fowls deposit over half their droppings at night, and the use of the droppings boards thus helps to keep the floor clean. The manure can then be easily collected, dried and stored in empty feed sacks for use as an excellent fertilizer for plants requiring organic nitrogen. The area under the droppings board then becomes an ideal site for a communal nest.

Providing Feed

In both intensive and semi-intensive systems, laying hens need constant access to food and water, and feeders should be distributed evenly throughout the chicken house. In the semi-intensive system, birds scavenge during the day, mostly for protein (from such sources as insects, worms and larva), minerals (from stones, grits and shells), and vitamins (from leafy greens, oil palm and nuts), but energy supplements such as maize, sorghum and millet are important for higher productivity and should be given.

Feeders

A good feeder should be:

— durable enough to withstand frequent cleaning;

— stable enough not to be knocked over;

— of the correct height and depth;

— bird proof (such that birds cannot get into it or roost in it); and

— equipped with a lip to prevent birds from spooning feed out onto the floor with their beaks.

The height of the feed inside the feeder, which should never be more than one-third full, should be level with the back of the birds, to prevent them from scratching contaminated litter into the feeders and to limit feed wastage. This is achieved by adjusting the height of the feeder itself. To reduce spoilage and mould problems, feed should be supplied at sunrise and at about 14.00 hours (or more frequently if the birds empty the feeder), with all feed finished by sundown.

Table 3. Feed and feeder space requirements for 100 chickens

Age(weeks)	Daily feed consumption (kg)	Suggested feeder depth (cm)	Feeder space (m)
1 - 4	1.4 - 5.0	5	2.5
4 - 6	3.2 - 7.3	8	3.8
6 - 9	5.0 - 9.5	9	6.1
10 - 14	7.3 - 15.9	12.5	9.6
15 and above	9.1 - 11.4	15	12.7

Feeders can be made of wood, sheet metal or bamboo, and are best suspended from the roof to keep rats out. The height of the feeder should be adjustable. Supplementary vegetable matter should be fed at beak level, either hanging from the ceiling wrapped in a string or placed in a net or placed in a floor-standing hopper with wire or slatted sides. It should not be thrown on the floor.

Feeder space is measured as the linear distance of lip available to the birds. This is either the circumference of a round tube-feeder tray or twice the length of a trough if the birds feed from both sides. If troughs are used, at least 10cm of feeding space should be accessible to each bird. When circular feeders are used, there should be at least 4cm feeding space per bird.

Creep Feeders

Creep feeders are used to enable baby chicks to have access (by "creeping" through a small doorway) to high-quality (high in energy and protein) feed, while blocking access to larger sized birds (especially the chicks' mothers). The conical (open at the top and base) creep feeder can be made from split bamboo strips approximately 0.5 to 1 cm wide, bound at the joints with string or wire. The base of the creep feeder can be 75 cm in diameter and 70 cm high, with an access hole (reinforced to form a carrying handle as well) about 20 cm wide at the top.

The gaps between the upright slats can be 2 to 3 cm at the bottom and about 1 cm at the top. The flexibility of the bamboo strips allows the size of the entryways to be enlarged, as the chicks grow bigger. If the chicks are reluctant to leave their mothers' side, then a more tightly formed weave can prevent their exit once they are placed inside the cone through an opening at the top. The bamboo can be preserved from insect attack with a coat of used engine oil.

Better nutrition for young stock boosts their immune response to disease challenge and to vaccine response by developing full immunity. Gunaratne et al reported that chick mortality rates could be reduced by the use of creep feeders but that this did not increase the growth rate. However, when the household waste was supplemented with protein and fed in the creep feeder, both the growth and the survival rate of chicks increased.

Table 4 shows that annual egg production can be doubled because of the increased laying time available to the hens if their chicks are fed in a creep feeder after hatching. If the gaps in the creep feeder are adjusted, it can also be used for growers over eight weeks, and if given less than the full ration, they will learn to compete for food with other chickens.

Table 4. The effect of creep feeding on flock egg production

Intervention	Period					
	June	July	Aug.	Sept.	Oct.	Nov.
Creep only	31.5	28.7	27.3	21.8	21.4	33.0
Creep + Low supplement	21.2	18.8	22.9	26.9	30.7	31.1
Creep + High supplement	24.3	24.5	32.5	34.1	27.4	31.1

Providing water

Providing clean water is a priority often neglected. The amount of water, the right type of equipment and where it is situated are important considerations. Table 5 shows water consumption rates for hot dry conditions, and these can be halved for temperate regions.

Table 5. Minimum water and watering space requirements for 100 birds in hot dry conditions

Age (weeks)	Daily consumption (litres)	Water space (m)
0 - 1	3	0.7
2 - 4	10	1.0
4 - 9	20	1.5
9 or more	25	2.0
Layer	50	2.5

In countries with plenty of water, such as Bangladesh, Viet Nam, Indonesia, the Gambia, Sierra Leone, Zaire and

Uganda, the only special measure to be taken is to ensure its cleanliness. In other regions, especially in the Sahel and other drought-prone regions, fetching and carrying water is a crucial task, usually assigned to women and children. In the rainy season, clean water and feed must be placed in the house, as brooding hens are kept inside to prevent poor hatchability resulting from contamination with mud and dirt.

The simplest equipment is a tin can inverted into a soup plate or the bottom of a larger tin can. A hole is punched about 2 cm from the open end of the tin can. The can is filled with water, covered with the plate and both quickly inverted. The position of the punched hole and the vacuum in the tin can will regulate the water level in the plate. Tin can waterers work well but quickly become rusty, especially in the humid tropics.

A clay pot or gourd with holes around the sides sunk into the ground for stability can be used to water adult birds. Clay pots of any dimension can often be ordered from local pot makers. Because they are permeable they provide water half a degree Celsius cooler than other waterers, as heat loss through evaporation keeps the water cooler. This also means an appreciable loss of water over time, especially in very hot dry areas, so that the pots may have to be made more impermeable by glazing. If continuous drinking troughs are used, at least 5 cm of trough should be allowed for each bird. Alternatively, one cup or nipple drinker may be provided for every ten birds.

Baby Chick Management

Baby chicks should be kept warm and dry. The nest, which they share at night with the mother hen, must be kept clean. In colder climates (below 20°C at night), the nest site should be kept warm by lining it with straw and placing it near a stove or fireplace. The chicks should remain with the mother hen for nine to ten weeks, learning from her example how to scavenge and evade predators and other dangers. Clean drinking water and fresh feed in a clean container should be provided to supplement scavenging.

There is a close relationship between chick weight and growth and mortality rates. In an experiment where young chickens had access to supplementary feed in a creep feeder, it was found that supplementary protein feed had a significant effect on the survival rate and growth rate. Chicks separated from the mother hens during the day from the age of three to ten weeks, and fed with chicken starter mash ad libitum, had a mortality rate of 20 percent and a body weight of 319 g at ten weeks, compared with a mortality rate of 30 percent and a body weight of 242 g for the control group which remained with the mother hens. A suitable strategy for rearing chicks therefore would be as follows:

- The chicks should be confined for the first weeks of life and provided with a balanced feed.
- A vaccination programme should be followed.
- Sufficient supplementary feed should be provided during the remaining rearing period to allow the chickens to develop in accordance with their genetic potential.
- Feed supplements and protection should be provided to naturally brooded chickens during the first four to eight weeks of life.

The composition of the supplementary feed will depend on the available scavengable feed, but a form of cafeteria free-choice feeding of a protein concentrate, energy concentrate and calcium mineral in each of three containers may be the best solution.

The mortality rate of naturally brooded chicks, whose only source of feed is from scavenging under free-range conditions, is very high and often exceeds 50 percent up to eight weeks of age. Wickramenratne et al found that predators accounted for up to 88 percent of mortality and that coloured birds had a higher survival rate than white birds.

The high mortality rate and the large number of eggs

required for hatching are the main causes of low offtake from scavenging poultry flocks. Smith reported an offtake (sales and consumption) of only 0.3 chickens per hen/year from a survey done on flocks in Nigeria. This low offtake has also been observed in Bangladesh and India.

An efficient way of decreasing mortality rate (a costly loss) is to confine and vaccinate the chicks during the rearing period. This however is more expensive, the cost of feed in particular increasing production costs. A method used over the past ten years in many poultry development projects in Bangladesh confines the chicks during the first eight weeks of life. They are fed approximately 2 kg each of balanced feed and thereafter kept under semi-scavenging conditions. At eight weeks of age, they are less susceptible to attacks by predators and more resistant to diseases, due to their larger body weight and more effective vaccination immunization (due to their better nutrient intake).

Hygiene

Manure management

Whatever the type of confinement, proper attention must be paid to manure management. Adult birds produce 500 g of fresh manure (70 percent moisture content) per year per kg of body weight. To preserve its fertilizer value, manure should be dried to about 10 to 12 percent moisture content before storage. This will retain the maximum nitrogen content for fertilizer value. Nitrogen in the form of urea is the most volatile component of manure, and is lost as ammonia if moisture content is too high in the stored material. If the moisture content is too high, then the stored manure releases ammonia, carbon dioxide, hydrogen sulphide and methane, which can have serious physiological effects on humans. Some of these components are also greenhouse gases, which contribute to the global increase in ambient temperature. Poultry manure is very useful as an organic fertilizer, as animal and fish feed and as a raw material for methane gas generation in biogas plants for cooking fuel.

Other hygiene management measures

Good ventilation discourages the spread of diseases and pests. In overnight houses, the provision of perches or loosely plaited bamboo mats (such as those used for sieving) placed on the floor can help to keep them dry. If the birds are housed inside, the floor should be swept daily. An outside chicken house should be cleaned every week to break the breeding cycle of the common housefly. It takes about seven days to complete the breeding cycle from fly egg to hatching of the adult housefly. Wood ash and sand spread on the floor will discourage lice infestation.

Mothballs (naphthalene) crushed with ash can also be applied to the feathers or the wings of the birds, or placed where the chickens usually take their dust baths. If the chickens are already infested with mites, the house can be fumigated (while the chickens are outside) with a rag drenched in kerosene. Lice live on the birds, and dust baths with naphthalene powder in the ash will be more effective than dust alone.

The practice of keeping chickens and ducks together should be discouraged. This results in wet floors, giving rise to diseases such as Fowl Cholera. Ducks are also much more tolerant than chickens to Newcastle Disease, and are thus often carriers of this viral disease. Adults and young stock of any poultry should be housed separately to minimize cross-infections and injuries from bullying.

Management of Free-range Poultry

The unrestricted free-ranging of poultry is often a problem. They trespass onto neighbouring fields and gardens, and are constantly at risk from predators. Confinement is often not practical because of the cost of feed and fencing, while surveillance is only feasible where the very old or very young of the household have time to help. Fencing of vegetable plots is in many cases the best option. Placing more cocks in the village might reduce the movements of the chickens, as the cocks and hens of each flock would keep more to their own territory. Cocks move within an eight-to-ten-house territory, and hens within two or three houses.

Under the free-range system, the difference between the amounts of food gathered through scavenging and the total food requirement for maximum production should be balanced with nutrients supplied from supplementary feed. To make up a properly balanced supplement, it is necessary to know the scavenger feed resource base (SFRB) and the composition of the crop contents.

If this is not known, it is recommended that the fowls have access (using a free-choice cafeteria system) to three containers (or three compartments of a bamboo stem feeder of ingredients comprising a protein concentrate, a carbohydrate source (for energy) and a mineral source (mainly for calcium carbonate for egg shell formation for the hen). Poultry should have free access to this cafeteria system for two to three hours in the evening to supplement the day's scavenging.

From a feed resource point of view, this recommendation is only economically viable (sustainable) if the consumption of supplementary feed per egg produced is equal to 150 to 180 g or less. Consumption of over 150 g is only justified if the supplements are cheaper than the commercial feed used in intensive poultry production. Supplements are usually recommended in the range of 50 to 80 g/bird/day, so it is usually quite viable. Seasonal variations in the SFRB have a substantial effect on production.

During the dry season, scavenged feed from gardens, crops and wasteland (such as grass shoots, seeds, worms, insects and snails) stops, while the quantity and quality of household kitchen waste decrease. The feed supplement should be adjusted seasonally to maintain an optimum level of production or, alternatively, the chicken population could be adjusted to the amount of the SFRB and the feed supplement.

Hens in confinement fed a balanced diet will convert food weight to egg weight at an efficiency of about 2.8 kg of feed per kilogram of egg weight. Changes in husbandry alone may increase the productivity of scavenging village chickens, without the need for additional inputs. In planted orchards, a stocking rate of 120 to 180 birds/ha will clean up windfalls while also fertilizing the trees. In this example, the amount of fertilizer produced per hectare for 150 hens (weighing two kilograms each) is based on the assumption of 500 g of fresh (70 percent moisture) weight of manure produced per kilogram of live weight per year.

This results in 330 g of dried manure (dried to a ten percent moisture content) per hen/year, and thus the 150 hens will produce 49.5 kg of dry manure per year. This has an equivalent fertiliser value of 13 percent ammonium nitrate, 8.6 percent super-phosphate and 2.9 percent potash (potassium) salts. Thus the 150 hens will produce the equivalent per hectare/year of 6.4 kg of ammonium nitrate, 4.3 kg of super-phosphate and 1.4 kg of potash salts.

Planning Flock Production and Size

Production involves birds for meat and eggs. For both meat and egg production, the number of chickens in the flock is the most important factor. Flock size changes constantly as eggs hatch and hens are sold or eaten. Usually the main cause of flock depletion is mortality, particularly in chicks. Disease is the greatest cause of mortality, especially in the rainy season and in the weather changeable humid periods on either side.

During summer and the rainy season, predators in the cropped fields also contribute to reduced flock sizes. Local birds lay an average of three to four clutches of 12 to 15 eggs in a year, with more eggs laid at crop harvest time because more feed is available. Given most traditional farming systems, keeping the flock number constant requires eight to ten eggs for reproduction, leaving an average of 35 to 40 eggs per layer for sale or consumption. Because the number of eggs needed for replacement may decrease with better management, the extra eggs can be sold or eaten. Most egg laying takes place between sunrise to mid-morning. During the months of laying, nest location should not be moved, as this may upset the laying routine.

In village flocks, income derives from the sale of eggs and live birds. For example, a flock of 15 local hens laying 30 eggs/hen/year (with one local cock) will produce 450 eggs in a year. Of these 450 eggs, 120 may be incubated by broody hens (in ten clutches of 12 eggs each), of which 100 chicks may hatch, and 30 eggs may be cracked and consumed in the household, leaving a balance of 300 eggs for sale. Of the 100 day-old chicks, 30 may reach maturity (with rearing losses of 70 percent), to yield 15 cockerels and 15 pullets. The 15 pullets will replace the older hens, of which ten remain after the sale of cull hens, and one new cockerel will replace the old cock. The annual income from the flock can therefore be calculated as follows:

300 eggs + 10 old hens + 1 old cock + 14 cockerels = income

For improved productivity, culling is important and productive birds should be carefully selected. For simplicity, the above example assumes no adult mortality.

Case Studies of Family Poultry Management Systems

Free-range System in Ghana

In the traditional free-range system of the Mamprusi tribe in northern Ghana, the farmer releases his 19 chickens and six guinea fowls from the space under the grain store each morning. Grains are thrown on the ground to feed the birds. A young boy takes care of the birds during the day and protects the crops from poultry damage. Occasionally the boy will feed the birds with a piece of termite hill, and in the evening he returns with the flock and locks them under the granary store.

Table 6. Total annual production of a Mamprusi average flock

	Chickens		Guinea fowls		Total Production
	Flock	Production	Flock	Production	
Cocks	3	1	2	1	2
Hens	9	3	4	2	5
Cockerels	2	22	-	13	35
Pullets	5	19	-	11	30
Eggs (/hen)	-	45 (20)	-	65 (50)	110 (380)

Chicken hens lay throughout the year, but guinea fowls lay only in the rainy season. Chicken hens produce about 20 to 40 eggs a year and guinea fowls about 50. Most of the eggs are used for hatching. Chickens also hatch guinea fowl eggs, as guinea fowls are not good mothers. Hatching takes place throughout the year, although most of the hens incubate their eggs in the rainy season. A reproduction cycle (laying, hatching, caring for chicks and resting) takes about 20 weeks. Mortality is high (75 percent) among the young chicks. Out of ten chicks, only about two reach adulthood, due mainly to disease, predators and road accidents. Newcastle Disease in particular kills many poultry in the dry season.

Worms as internal parasites are a great problem, weakening the birds. Predators include snakes, birds of prey, cats and dogs. Mortality up to two months of age is 50 percent, with a further 25 percent thereafter up to sexual maturity. Hatchability of guinea fowl eggs is very low (45 percent) compared with chickens (72 percent). Farmers keep hens for about three years and guinea fowls for two years, after which productivity decreases considerably and they are culled.

Scavenging Commercial Hybrid Layers in Sri Lanka

In a study carried out by Roberts and Senaratne, Sri Lankan villagers reared hybrid egg layers in a semi-scavenging system. Day-old hybrid chicks were brooded under the heat of a small kerosene lamp. The chicks were provided with a little mixed supplement of local crop by-products, comprising

40 percent rice polish, 50 percent expeller coconut meal and 10 percent broken rice. The Proximate chemical analysis of this supplement was 16 percent Crude Protein, 8 percent Crude Fat, 7 percent Crude Fibre and 7 percent Ash.

The amount of the supplement increased from 8 to 60 g/bird/day until 12 weeks of age, and was maintained at 60 g thereafter. The growth rate was 38 g/bird/day up to 20 weeks of age. The mortality rate of the chicks was only four percent in the period up to ten weeks, which compared favourably with mortality of 68 percent up to six weeks in Indonesia and 25 percent up to eight weeks in Thailand in chicks hatched and reared by village hens. The comparative advantage of the Sri Lankan performance was attributed to supplementing the competitive scavenging, and to the protection against predators provided by the semi-intensive management system. It is probable that chicks would also benefit from the use of a simple creep feeder for feeding kitchen waste.

The mortality rate, in the Sri Lankan example, increased after reaching eight months of age, perhaps due to a greater need for scavenger free-ranging, and almost reached a cumulated 60 percent loss by 13 months of age. Of the 142 hens lost up to 13 months of age, records were kept for 92. The causes of mortality were:

- 32 percent predators (such as dogs, mongooses, pole cats and snakes);
- 26 percent disappeared;
- 15 percent Newcastle disease;
- 15 percent intestinal infection;
- 5 percent stolen;
- 4 percent accidents (vehicles and falling coconuts); and
- 2 percent attacked by humans.

Hens laid their first eggs when they reached 21 weeks (146 days) of age, although 40 percent production (on a hen/day basis) was not achieved until they were 30 weeks of age. Peak egg production was just over 60 percent. A severe drop in production (beginning when the hens were eight months of age) corresponded with an outbreak of Newcastle Disease in local village birds and the start of the long dry intermonsoon period. Production fell to below 30 percent when the hens were ten months of age, and slowly rose again to over 60 percent at 13 months of age. The recovery in production began during the dry period and was maintained into the next season. Egg production was comparable with that of hybrid egg layers, which were introduced into the village as pullets, provided with a supplement and allowed to scavenge.

The production was much better than the 12 to 21 percent reported in village birds in Indonesia and in Thailand. The egg weight reported by Roberts and Senaratne was 60 g compared with about 40 g for village hens.

A Free-range System in Senegal

In a study carried out on farms in Senegal, flock sizes ranged from under five birds to more than 15 birds, with an average flock size of ten birds. Seven percent of the flocks comprised under five birds, 38 percent comprised five to ten birds, 41 percent comprised 10 to 15 birds, and 14 percent comprised more than 15 birds.

Flocks with fewer than five birds had either recently lost hens or had hens that had not yet hatched their eggs. Flock size varied considerably during the year, due to additions (hatchings, purchases and gifts) and to chickens either sold or lost through disease or predators. The birds were permitted to scavenge during the day and were locked into wooden cages (ngounou) at night for protection. The cages were made on the farm from available materials (including bricks, galvanised iron sheets and wood). The doors were small, to prevent entry by thieves and predators. Stock density in the cages was about 25 birds/m^2. Feed and water was available to supplement kitchen waste and scavenging.

Table 7. Village flock structure in Senegal Source

Age(months)	N°	Males	N°	Females	N°	Total%
0 - 1	-	-	-	-	320	50.5
1 - 3	-	-	-	-	99	15.6
3 - 6	34	5.4	84	13.3	118	18.6
6 - 8	2	0.3	21	3.3	23	3.6
8 -10	1	0.2	19	2.9	20	3.2
10 +	15	2.4	39	6.2	54	8.5
Total	52	8.3	163	25.6	634	100

Table 8. Age-related mortality in local birds in Senegal

Age(weeks)	% Mortality(cumulative)
1	13 ± 5
2	15 ± 9
3	39 ± 20
4	42 ± 20
8	49 ± 20
12	66 ± 17

The proportion of young chicks and growers in the flock was about 60 percent while adults represented 40 percent. Mortality in the first month of age was 40 percent. There were four to five clutches of eggs laid per year, with 8 to 15 eggs per clutch. Egg weights ranged between 38 and 43 g with an average of 40 g. Almost all eggs were set for hatching and of these hatchability was about 80 percent.

Table 9. Body weight of local birds in Senegal

Age	N°	Males	N°	Females	N°	Males+ Females
1 week	-	-	-	-	81	34±5
2 weeks	-	-	-	-	75	58±10
3 weeks	-	-	-	-	66	101±43
1 month	-	-	-	-	98	171±70
2 months	-	-	-	-	41	464±242
3 months	-	-	-	-	58	631±211
4 months	29	975±20	63	746±170	92	860

6 months	5	1380±150	21	1229±165	26	1305
8 months	2	1826±75	21	1264±183	23	1544
10 months	1	1500	19	1245±150	20	1372
+ 1 year	15	1803±4	39	1350±223	54	1577

The production cycle was eight to ten weeks (10 to 15 days for egg laying, 21 days for incubation, and only 34 days for rearing). The chicks remained close to the hens for up to two weeks, during which time there was a relatively low mortality rate of 14 percent. On leaving the immediate protection of the hens, mortality increased sharply to 40 percent between three and four weeks, and up to 66 percent by three months of age. Similarly, the average daily live-weight gain of birds under this extensive system decreased from 10 g at eight weeks to 6 g at 12 weeks.

References

Calnek, B.W., Barnes, H.J, Beard, C.W., Reid, W.M. & Yoderg H.W.Jr., *Diseases of Poultry,* Ninth Edition, Iowa State University Press, Ames, U.S.A., 1991.

Leeson, S. & Summers, J.D., *Commercial poultry nutrition,* University Books, Guelph, Canada, 1991.

Merritt, E.S. & Clarridge, R.W.E., "The effect of length of holding on the hatchability of goose eggs", *Poultry Science,* 1959.

Rosinski A., Rouvier R., Wezyk S. & Sellier N., "Reproductive performance on geese kept in different management systems", *Preliminary Proceedings, 10th European Symposium on Waterfowl,* World's Poultry Science Association, 18-25, Halle, Germany.

Rosinski A., Rouvier R., Guy G., Rousselot-Pailley D., & Bielinski A., "Possibilities of increasing reproductive performance and meat production in geese", *Proceedings, XX World's Poultry Congress,* New Delhi, India, 1996.

Rosinski A. & Wencek E., "Current problems on waterfowl breeding and production (in Polish)", *Proceedings of Symposium "Poultry-Farm",* Gdynia, Poland, 1997.

10

Poultry Production: Low Cost Technology

Simple Shelters for Native Chickens

Native chickens have a number of useful characteristics, including the ability to thrive on coarse feeds, resistance to diseases, and lower nutrient requirements. A framework of a simple shed can be made of bamboo cane or tree trunks, with a roof of rice straw or lalang grass (*Imperata cylindrica*). Chickens can rest at night and on rainy days under shelter. A cardboard box, with dried weeds or rice straw inside, can be put under a tree, about 20cm from the ground, as a nest. Discarded containers such as tin cookie boxes with a stone inside are used as feed troughs and water containers. Feed troughs can also be made of wood.

The feed and water troughs are placed in the shade, and adjusted to the height of the chickens. The number of nests, feed, and water troughs should be sufficient. It is better to choose a feeding place where there is good drainage.

In addition, shrubs and weeds around the feeding area will provide places for chickens to rest and look for feed. Grain and by-products should be supplied when feed is not sufficient. It is better to fence the feeding area with bamboo cane, rice straw or Lalang grass. In places where electric power for an incubator is not available, or in less developed areas where there is a shortage of vaccines and commercial feeds, farmers should choose native chickens that brood, are climate-tolerant and disease resistant. It will be easier to get a good number of chicks and a high rearing rate.

Chicken houses can be constructed on barren land. Insect, weeds, leaves, kitchen leftovers, broken grain, and grain by-products can be used as feed. Chicken production of this kind not only improves the fertility of the land but also minimizes production costs. It gives local residents a supply of animal protein.

Farmers using simple shelters for native chickens should pay attention to the repair of the shed before the rainy season, to prevent the chickens from getting wet. In winter, it is better to build a fence made of rice straw or bamboo, etc., to protect the chickens from wind. Litter in the nest should be kept clean and replenished as necessary.

Agricultural by-Products As Floor Litter for Ducks

By-products such as rice hull, corncobs, peanut hulls, and sugarcane bagasse can be used as floor litter in duck production. They are commonly placed inside the sheds where the birds feed and lay eggs. The litter is spread out to cover the ground with a layer about 3-4 inches thick. As the litter becomes soggy with moisture after several days, it is removed and additional litter is spread.

Crop residues used as floor litter keep the living environment of the birds dry and help prevent the spread of pests and diseases. The litter materials also serve to cushion the eggs, minimize breakage and keep the ground covered. Thus, the litter promotes the collection of clean eggs, which command a higher price. Also, the used litter when collected can be spread in the field as organic fertilizer.

Once the litter becomes soggy with water, immediate replacement is needed. Otherwise it becomes a very effective medium for the spread of diseases. Disposal of used litter can also be a problem. With the use of litter, some eggs may be left uncollected because birds have the tendency to cover their eggs once laid.

Cage for Native Chickens

Strips of bamboo approximately 1cm wide and 0.5cm thick are used to construct a hemispherical net shape with a large hole in the top.

The fundamental behaviour of chickens (scratching for food) is not restricted. The cage is usually used to keep a fighting rooster in, or a hen with its young brood. The chicks are free to run in and out of the cage. The bamboo cage is placed on the ground with the chickens inside, so that the chickens are able to scratch on the ground. Mating of specific individual chickens can be controlled when the cage is used.

Feed and water must be provided at all times. The large hole at the top of the cage must be plugged with a discarded bowl with a brick in it.

Pandan Bags for Individual Transport of Poultry

Avery popular indigenous material for weaving handicrafts like handbags, mats, transport bags and hats is pandan (family Pandanaceae, genus Pandanus). Also known as crew pine, pandan is a tropical shrub found in the Philippines that varies from one meter to more than 15m in height. This plant grows in backyards, along coastal shores, and in elevated areas. In moist places, the plants grow naturally under coconut.

Immediately after harvest, the pandan leaves are cut and stripped to the desired size then sun-dried for 3 days. The leaves are pre-pressed to improve the quality and prevent wastage. A heavy round log is commonly used as a press. The flattened leaves are then woven into coarse or fine baskets or bags.

The pandan bag is commonly used for transporting individual native chickens, especially fighting cocks. However, other poultry such as ducks, geese, and fowls can also be transported using this bag. Holes are made on both sides of the basket to allow ventilation. The bags are hand-carried, or can be loaded onto trucks, boats, etc. When the bags are well woven and sturdy, they can carry relatively heavy weights. Containers of this kind are also used to carry vegetables, fruits, meat, fish, etc. The bag can be re-used, and lasts a long time it stored properly.

As a means of carrying and transporting poultry like chicken, game fowls, ducks, etc. the pandan handbag or basket is quite popular. One person can easily carry several bags. It is durable, and at the same time cool, since it is made of natural materials. The bag is also cheaper than conventional boxes. The pandan bag can allow only one mature bird per bag. Being made of light material, the bag can easily be trampled on, endangering the bird inside.

Perch for Native Chickens

Two wooden posts are secured to the ground and a piece of bamboo is fastened to them. A small ladder is then constructed and then fastened to the bamboo.

Native chickens still retain some original behaviour of wild fowl. They perch at night on branches of trees or bamboo. The perches are usually placed inside a chicken house. The native chicken will climb the bamboo ladder and perch on the long piece of bamboo for the night. This will improve the hygiene of the chicken as well as provide some protection from predators such as snakes.

The perches should be placed on one side of the chicken house. The space on the perch should be sufficient for all chickens to prevent competition for space at dusk. Sharp splints and projections on the bamboo should all be removed to prevent injuries.

References

Aklobess, K.K., "Smallholder rural poultry production in Togo", In *Proceedings. CTA Seminaron Smallholder Rural Poultry Production, Thessaloniki. Greece 9–13 October* 1990.

Alders, R.G., "Women and village poultry: an old tradition returns to the fore", *African Network for Rural Poultry. Newsletter*, 1997.

Chaheuf, N., "Disease prevention in smallholder village poultry production in Africa", In *Proceedings, CTA Seminar on Smallholder Rural Poultry Production*, Thessaloniki. Greece, 9–13 October 1990.

11

Natural Incubation of Poultry

The broody hen chosen for natural incubation should be large (to cover and thus keep more eggs warm), healthy and preferably vaccinated, with a good brooding and mothering record. Signs of broodiness are that the hen stops laying, remains sitting on her eggs, ruffles her feathers, spreads her wings and makes a distinctive clucking sound. Brooding may be induced with dummy eggs or even stones.

Eggs usually become fertile about four days after the rooster has been introduced to the hens. A maximum of 14 to 16 eggs may be brooded in one nest, but hatchability often declines with more than ten eggs, depending on the size of the hen. Feed and water provided in close proximity to the hen will keep her in better condition and reduce embryo damage due to the cooling of the eggs if she has to leave the nest to scavenge for food.

The hen keeps the eggs at the correct humidity by splashing water on them from her beak. This is a further reason for providing her with easy access to water. In very dry regions, slightly damp soil can be placed under the nesting material to assist the hen in maintaining the correct humidity (between 60 and 80 percent). Fertile eggs from other birds are best added under the brooding hen between one and four days after the start of brooding. In Bangladesh, it has been reported that local broody hens will even sit on and hatch a second clutch of eggs, often losing considerable weight in the process.

The incubation period for chicken eggs is 20 to 21 days, and increases up to 30 days for other poultry. After sitting for some days, a broody hen can be given some newly hatched chicks and, if they are accepted, the original eggs can be removed and replaced with more chicks. Thus hens with a better record of mothering can be better utilised for their abilities.

Eggs initially need a very controlled heat input to maintain the optimum temperature of 38°C, because the embryo is microscopic in size. As the embryo grows in size (especially after 18 days), it produces more heat than it requires and may even need cooling. Moisture levels of 60 to 80 percent Relative Humidity (increasing during the incubation period) are important to stop excess moisture loss from the egg contents through the porous egg shell and membranes. Factors to consider for successful natural incubation include the following:

— Feed and water should be close to the hen.

— The broody hen should be examined to ensure that she has no external parasites.

— Any eggs stored for incubation should be kept at a temperature between 12 and 14 °C, at a high humidity of between 75 to 85 percent, and stored for no longer than seven days.

— Extra fertile eggs introduced under the hen from elsewhere should be introduced at dusk.

— The eggs should be tested for fertility after one week by holding them up to a bright light (a candling box works best. If there is a dark shape inside the egg (the developing embryo), then it is fertile. A completely clear (translucent) egg is infertile.

A hatchability of 80 percent (of eggs set) from natural incubation is normal, but a range of 75 to 80 percent is considered satisfactory. Setting of hatchings is best timed so that the chicks to be hatched are two months of age at the onset of major weather changes, such as either the rainy (or dry) season or winter/summer. A plentiful natural food supply over the growing period of the chicks will ensure a better chance for their survival. Successful poultry species instinctively lay and incubate their eggs at a time of the year when newly hatched chicks will have a better supply of high protein and energy food provided by the environment. For example, guinea fowl will only lay eggs in the rainy season. However, seasonal changes in weather patterns are also times of greater disease risk.

Artificial Incubation

There are many commercial artificial incubators of varying capacities. Most depend on electricity, but some use gas or kerosene for heating. All use a thermostatic switching device to keep the temperature constant within one Celsius degree. The correct humidity is usually maintained by having a predetermined surface area of water appropriate for each incubator chamber.

Turning the egg several times each day is important to prevent the embryo from sticking to the shell membranes. With hand-turning systems, an odd number of times turned per day (five to seven times) will ensure that during successive overnight periods the egg is always oriented the opposite way from that of the previous night.

The broody hen carries out all of these incubation tasks instinctively, and artificial incubation attempts to duplicate these tasks. Traditional artificial incubation techniques have evolved over thousands of years in many parts of the world. One such technique, developed for hatching duck eggs in China, is the parched (heated) rice technique. It is based on the use of heated paddy rice and embryo-generated heat. It is still used in China and Bali, Indonesia, with hatchability results of up to 80 percent. The objectives of artificial incubation are met equally well using either parched rice or rice husks, and a hatchability of 65 to 75 percent is common. By candling the eggs between days 5 and 7, infertile eggs can be detected as "clears" (as the light is not obscured by the growing embryo). These eggs are still suitable for sale for human consumption, which improves the economic viability of this system.

As duck eggshells are less brittle than chicken eggs, the system was never adopted for chicken eggs in China. The original Chinese system used 80 duck eggs per bundle. However, with extra care, and fewer eggs per bundle (25 to 30 compared with 40 duck eggs), chicken egg incubation was found to be equally successful in Bangladesh when adapted there in the 1980s. The number of duck eggs per bundle was reduced to 40, which gave better hatching results as well as fewer breakages than 80 per bundle.

The artificial parched rice or rice husk incubation process is started by heating the eggs, either in the sun or in an insulated warming room equipped with a heat source. On sunny days, approximately 25 to 30 chicken eggs or 40 duck eggs, (presumed fertile, and carefully dated and labelled) are placed in the sun on pieces of padded cloth for about 30 minutes and turned occasionally to raise the temperature of the eggs to the required 37 to 38 °C. This temperature can be judged by the appearance of water droplets on the shell or by touching the egg to the eyelid.

On sunless days, eggs must be placed on a cloth in a shallow bamboo basket and put on racks in a heated warming room to slowly achieve the same temperature. This usually requires approximately one to three hours. Any slow-burning fuel is suitable, and kerosene and charcoal are commonly used. In Egypt, dried beanstalks were used for thousands of years for their characteristic slow-burning property, and are still used today in the Fayoumi district (from where the well-known chicken breed of the same name originates). A well-vented stove will prevent any toxic fumes affecting the embryo.

In Bangladesh, slightly heated (to 38°C) sand or wood-ash, covering the eggs for approximately one minute, is also effective in warming the eggs. Human clinical thermometers, now readily available, can be used to assist training in using the eyelid's sensitivity to temperature. If the humidity drops below 70 percent, 60 percent, 50 percent or even 40 percent (particularly in typical progressively drier months), a wet (and slightly warm) cloth should be placed over the eggs with a frequency corresponding to the above humidities of one, two, four or six times daily. This will raise the humidity of the egg so that the embryo will not dry out.

The unhusked (paddy) rice is heated (parched) and continuously stirred until it reaches a temperature of 60°C, to provide heat for the eggs for the first two weeks of incubation. About 2.5 to 3.0 kg of the heated rice is enclosed in a cloth pillow and placed into the egg basket. The pillows have the same diameter (40 cm in the Bangladesh example) as the basket, and should be about 8 cm thick. Where rice-husk is used, pillows can be made with black-coloured material, which easily absorb the sun's heat.

When the temperature of the pillow has dropped to about 40°C, a loosely bunched bundle of 40 duck (or 25 to 30 chicken) pre-warmed eggs are placed on top of the warm rice. The bundle is made from a square piece of cloth about 45 cm on each side. A soft duster cloth with pinholes is suitable. Alternating pillows of warm rice and egg bundles are added until the basket is full, finishing with a pillow of rice. The basket is then covered with padding to conserve as much heat as possible.

This procedure is repeated until all the eggs are placed in baskets, leaving one basket empty to allow the addition of freshly warmed rice. The incubating baskets are cylindrical in shape (50 cm in diameter and 80 cm in depth in China, and 40 cm in diameter and 70 cm in depth in Bangladesh and are made of woven bamboo strips. In China, a few layers of bamboo paper are pasted on the inner surface of the sides and bottom to seal the gaps against temperature loss through convection.

In warmer parts of southern China, rice husk is substituted for paddy rice, and the pillows containing the husks are made from black-coloured material, which easily heat up in the sun.

Rice husks also provide very good insulation against loss of heat from the older eggs, which is instead transferred (by conduction) to younger eggs placed in contact with them in separate bundles. This rice-husk system was adopted on a large scale in Bangladesh after being introduced by poultry development projects in the 1980s.

The system has evolved, and the cylindrical egg baskets are now set into larger bamboo frame setting boxes, with more insulating rice husk material placed between the cylinders and the walls of the enclosing setting boxes. The cylinder wall should be about 10 cm from the setting box wall and 8 cm from the next cylinder. With this greater insulation, there is less heat loss, thus less need to provide supplementary heat from costly fuels.

For the first three days, reheated paddy rice (or rice husk) is added three times a day at regular intervals. During days four to six, this may be reduced to twice a day. The object is to ensure that the eggs are kept at the temperature most suitable for embryo development. The spare basket is used to transfer eggs from an adjoining basket when adding freshly warmed rice or rice-husks. Thus the top layer of eggs becomes the bottom layer and the bottom layer ends up on top of the spare cylinder. The newly emptied basket is then ready to receive eggs from the third basket, and so the cycle continues.

In China, heat is provided for days 13 to 14 in summer and in the colder months also on days 18 to 19. After that, the developing embryos are able to produce enough heat to maintain the incubation process without further need of an outside heat source. Eggs set for the first six days are called "new eggs", those between days 6 to 13 (which neither need nor produce extra heat) are "in-between eggs" and eggs after day 13 (which give out excess heat) are "old eggs".

Once a basket contains "old" and "in-between eggs", it is possible to use embryo-generated heat alone to incubate "new eggs", which are usually introduced at intervals into the basket and placed between layers of "old eggs".

Table 1. Arrangement of eggs in the incubation baskets prior to and after transfer to adjoining baskets

Layer	Before transfer	Incubation time (days) Summer	Winter	After transfer
1	A	14-6	17-20	J
2	B	1-3	1-4	I
3	C	10-13	13-16	H
4	D	4-6	5-8	G
5	E	7-9	9-12	F
6	F	7-9	9-12	E
7	G	4-6	5-8	D
8	H	10-13	13-16	C
9	I	1-3	1-4	B
10	J	14-16	17-20	A

Even when outside heat is not added, the eggs in the baskets must be regularly turned and aired three to five times a day in the process of transferring them to adjoining baskets. In order to use the embryo-generated heat effectively (almost all year round), the layers of eggs in the baskets are organised to a particular pattern, a typical example of which is shown in Table 1.

The eggs are candled on days 5 and 13, both to identify infertile eggs and dead embryos and to assess the degree of embryo development; which is used as one of the guides in adjusting basket temperature. Placing the egg on the upper eyelid allows the egg temperature to be assessed. The temperature of the basket may be adjusted in the following ways:

— by varying the proportion of eggs in a basket at different stages of incubation (for example, the temperature may be lowered by removing some of the bundles of "old eggs");

— by varying the arrangement of eggs in a basket (for example, as heat dissipation from eggs near the sides of the basket is faster than from those in the centre; a bundle of "old eggs" can be placed in the centre core and bundles of "new eggs" shaped to enclose them); and

— by changing the top covering of the basket using heavier padded material in cold weather and at the beginning of the incubation period, and a lighter covering if less heat retention is required.

Eggs in the advanced stages of incubation produce a lot of heat, so on days 13 to 14 in summer (days 18 to 19 in winter), the "old eggs" are transferred to hatching beds, where they are placed in a single layer for final development and hatching. The surface of the bed is covered with a thin layer of rice husks and then covered with a straw mat. The edges of the bed are lined with padding to protect the eggs.

The covering for the developing eggs in the bed may be heavy or light cloth, depending on the degree of insulation required. The temperature in the hatching bed is maintained at 36 to 37 °C, slightly lower than that of the basket. The temperature can be adjusted by changing the thickness of the covering, varying the space between the eggs, and moving the eggs twice a day so that those on the perimeter change places with those at the centre. In very hot dry weather, the eggs are sprayed with a fine mist of water. They are kept in the bed until the chicks hatch out and dry.

The hatching bed should be two-storied (like a bed-bunk) and can be made of wood. An example hatching bed with a 500-egg capacity has a length of 90 cm and width of 68 cm. The height of the side wall on all sides should be 20 cm from the bed base level, with a 25 cm gap above the side wall for ventilation and ease of access.

References

Bogenfürst, F., "The current state and future of incubation in waterfowl", World's Poultry Science Association, Preliminary Proceedings, 10th European Symposium on Waterfowl, Halle, Germany, 1995.

Brown, Edward, *Poultry Breeding & Production,* London, 1929.

Johnson G, and Wingfield W. W., *The Poultry Book,* London, 1853.

12

Managing Poultry Diseases

In most family flocks, disease is an important problem. Although farmers are familiar with the signs and symptoms of disease, the underlying causes are less well known. Almost every farmer and most extension workers hold Newcastle Disease (ND) responsible for most deaths, and the disease has a local name in all languages.

Table 1. Signs of poultry disease observed by farmers in East Kalimantan villages, Indonesia

Signs	Frequency %
Chickens huddle together	16.1
Coughing, sneezing, rapid breathing	13.2
Discharge from mouth and nostrils	10.9
Dullness, no appetite, closed eyes	10.9
Paralysis of legs and wings	9.2
White droppings	8.6
Turned or twisted neck	8.0
Dark red colour of head and comb	6.9
Greenish or yellow droppings	4.6
Bloody reddish droppings	4.0
Swellings of head and comb	2.9
Pale comb	1.7
Worms in faeces	1.7
Eye worm	1.1

However, not all infectious diseases are due to the Newcastle Disease Virus (NDV). Digestive problems resulting in slow growth and diarrhoea may be the result of rancid feed or too much salt, and may also be symptoms of diseases such as coccidiosis, salmonellosis, or Gumboro Disease (Also called Infectious Bursal Disease [IBD]). ND often has the symptom of greenish faeces, which indicates a loss of appetite.

Poultry have highly developed and sophisticated respiratory systems. Their heart and breathing rate are faster than humans, and their body temperature is 5°C higher. Their lungs are connected at the lower ends to a complex series of membrane-enclosed air sacs, which in turn are connected to air cavities in their major skeletal bones. These features contribute to their lightness and flying ability. The disadvantage is extreme susceptibility to respiratory infections caused by a wide range of bacteria, viruses and fungi. These infections become more of a problem in domestication, which usually involves some degree of increase in stock density - even if only for overnight accommodation - and thus increases the risk of cross-infection.

Inadequate ventilation of poultry houses results in a build-up of ammonia gas from poultry faeces, which contain urea. This can predispose the poultry to respiratory disorders, such as sneezing, running eyes and mucous discharges from the mouth. Providing good ventilation easily prevents this. More prolonged respiratory disorders are usually caused by diseases such as ND, Infectious Bronchitis (IB), Infectious Laryngotracheitis (ILT), Chronic Respiratory Disease (CRD) and diphtheria.

Coordination disorders such as paralysis, limping, twisted neck and slow movement may be caused by a variety of factors, such as physical injury, nutrient deficiencies and diseases, including ND (twisted necks or torticolis), Marek's Disease (paralysis), synovitis (tendon infections in which feet joints feel warm) and Avian Encephalomyelitis (AE).

Common Diseases

The common diseases and disorders of free-range poultry may

be either infectious or noninfectious, and are caused by a wide range of organisms or deficiencies.

Table 2. Causes and examples of poultry diseases

Causal Agent	Example
Infectious	
Virus	Newcastle Disease, Avian Encephalomyelitis, Fowl Pox, Marek's Disease, Infectious Bronchitis Infectious Laryngotracheitis, Gumboro Disease (Infectious Bursal Disease), Duck Virus Hepatitis
Mycoplasma	Chronic Respiratory Disease
Bacteria	Fowl Cholera, Salmonellosis, Pullorum, Fowl
Typhoid,	Infectious Sinusitis, Colibacilosis
Parasites	*Ectoparasites*: lice, mites, ticks
	Endoparasites: nematodes, Histomoniasis, Haemoparasites, round worms, hair worms, Avian
Malaria	
	Protozoa: Coccidiosis, Blackhead
Fungus	*Aspergillosis*: A. flavis (toxins), A. fumigatus (airsaculitis)
Non-Infectious	
Deficiencies	rickets, curled toe paralysis, encephalomalacia
Toxicities	salt poisoning, food poisoning (Botulism
Clostridium	botulinum and C. perfringens), poisonous plants

Infectious Diseases

Viral Diseases

Viral diseases are some of the most important infectious diseases affecting poultry. They are characterised by not being able to be treated, but most can be prevented with vaccines. The more important viral diseases are outlined below.

Newcastle Disease (ND)

This disease (called Ranikhet Disease in Asia) spreads rapidly via airborne droplets spread by the coughing or sneezing of infected birds. The virus can be carried by wild birds, through contaminated eggs, and on clothing. As mortality is often 100 percent in young chickens, ND is probably the most important constraint to family poultry development. Birds of any age can be affected, although young ones are more susceptible. Mortality in older chickens is usually lower, but egg production is usually severely reduced.

The incubation period of three to five days is followed by dullness, coughing, sneezing and gasping. Rapid breathing is accompanied by a gurgling noise in the throat. The respiratory signs usually develop first and are sometimes followed by nervous signs, characterised by twisting of the neck, sometimes combined with dragging of wings and legs. Depending on the environment and the degree of resistance of the birds, not all symptoms may be shown, or they may be in a mild or subclinical form. Some farmers have observed that the twisting of the neck occurs only in birds that survive.

Early loss of appetite results in a greenish diarrhoea. The most obvious diagnostic sign of ND is very sudden, very high mortality, often with few symptoms having had time to develop. Diagnosis of ND can be difficult from just the symptoms, as they are so varied, and as many other diseases share the same symptoms. The high incidence of ND among family free-range flocks is due to the following factors:

- the prevalence of virulent strains (velogenic, viscerotropic and pneumotropic) in tropical countries;
- continuous contact with other domestic and wild species of birds (such as ducks and pigeons), which can carry the virus without showing the disease (Majiyagbe and Nawathe, 1981); and
- uncontrolled movement of birds between villages.

There is a seasonal pattern to outbreaks of ND, influenced by:

- the arrival of migratory birds;
- changes in climatic conditions leading to stress, which predisposes birds to the disease;
- hot, dry and windy periods, which encourage airborne spread of the virus; and

— overuse of the few supply points of water available (during the dry season), which then become heavily contaminated with the virus.

Fowl pox

Fowl pox is still prevalent in many poultry flocks, for the following reasons:

— The fowl pox virus can remain alive in the pox scabs (which have fallen off the birds) for up to ten years, which contaminate the environment.

— Mosquitoes and other blood-sucking insects can transmit the virus.

The disease tends to be seasonal, occurring after mosquito breeding times. It is endemic in Papua New Guinea, where it is significant economically because the only NDV in the country is the non-symptomatic form. It is also a major disease in many other tropical countries.

Marek's Disease

Infection occurs early in life, and once a bird is infected, it can shed the virus in skin flakes throughout its life, if it survives. Clinical signs occur in young growing birds in the Acute Marek's Disease (MD) form, characterised by high mortality from visceral tumours. Another peak of mortality occurs in the Classical MD form, characterised by nerve paralysis in the legs and wings of birds aged from 15 weeks to early in the laying period.

Mycoplasmal Diseases

Mycoplasmas are not classified as bacteria or viruses, but as Pleuro-pneumonia-cocci-like organisms (PPLO). These are primarily associated with Chronic Respiratory Disease (CRD), a complex syndrome caused by Mycoplasma gallisepticum in partnership with bacteria (often E. coli), fungi and viruses (often Infectious Bronchitis). M.gallisepticum can be transmitted through the egg. Multi-age flocks, nutritional deficiency and water deprivation are important factors in the epidemiology of the disease in rural poultry flocks.

Bacterial Diseases

Fowl Cholera (Avian Pasteurellosis)

This is a contagious septicaemia (caused by Pasteurella multocida) that affects all types of fowls. It is often transmitted by wild birds or other domestic birds, and spreads by contamination of the feed or water and by oral or nasal discharges from infected birds. The incubation period is four to nine days, but acute outbreaks can occur within two days of infection.

In some cases, birds die within a few hours of showing the first signs, which vary depending on the form of the disease. The respiratory form is characterised by gasping, coughing and sneezing, while in the septicaemic form there is diarrhoea with wet grey, yellow, or green droppings. In the localised form, the signs are lameness and swelling of legs or wing joints. In acute cases, the head and comb change colour to dark red or purple.

If the infection is localised in the region of the ears, a twisted neck (torticolis) can sometimes be observed. In chronic cases, the comb is usually pale, with swellings around the eyes and a discharge from the beak or nostril. Fowl Cholera is common everywhere among free-range village flocks, because they are comprised of different species and are in continuous contact with wild birds.

Pullorum (Bacillary White Diarrhoea)

This is an egg-transmitted disease (caused by Salmonella pullorum) that spreads during incubation or just after hatching. White diarrhoea can be seen from three days to several weeks of age. The chicks refuse to eat, keep their heads tucked in and their wings hanging down. They huddle together and make a peeping sound. Mortality in the acute form ranges from 20 to 80 percent, and in the chronic form is around five percent. In the chronic form, the signs are a marked swelling of the hock joints, poor feather development, lack of appetite and depression. Table 3 shows the results of a survey in Zaria, Nigeria, conducted by Adesiyun et al. for Pullorum antibodies, which indicate a past infection with the bacteria.

Table 3. Prevalence of Salmonella pullorum antibody in chickens in northern Nigeria

Management	Age (wks)	N° tested	N° positive	(%)
Free-range	young	59	15	25
(indigenous)	adult	101	40	40
Backyard	young	90	8	9
(exotic)	20 +	70	24	34
Confinement	young	70	22	31
(exotic)	20 +	90	69	77

The free-range stock sampled was indigenous, and the other two groups were exotic. As the age of the birds in the free-range survey was not known, birds not in lay were counted as young.

Fowl Typhoid

Fowl typhoid is caused by Salmonella gallinarum, and commonly affects adult fowls. When it occurs in young birds, the signs are similar to those of S. pullorum. The incubation period is four to five days, and two days later the birds become depressed and anorexic. The colour of the comb and wattles becomes dark red; the droppings become yellow and the birds close their eyes and keep their heads down. Usually the affected chickens die within three to six days. Pullorum and fowl typhoid complex are both prevalent under free-range conditions.

Avian Salmonellosis (Paratyphoid)

Salmonellosis is usually used to describe infection with any organism of the Salmonella group other than S. pullorum or S. gallinarium. In countries with intensive poultry systems, poultry meat and eggs are a major source of infection for humans. The opposite may be true of family poultry, with humans infecting poultry.

Parasitic Diseases

External Parasites (ectoparasites)

These are very common in scavenging poultry, and include:

- *Lice*: these live on the skin of the birds, especially around the cloacae and under the wings. The irritation they cause can lead to reduced production. Lice species commonly found on poultry are Menacanthus straminens, Lipeurus caponis, Monopon gallinae, Goniodes gigas and Chelopistes meleagride.
- *Mites*: these are troublesome ectoparasites, which hide in the cracks of housing and perches, and come out only at night. They are bloodsuckers and lower egg production. Mites such as Dermanyssus gallinae can also transmit the bacteria Borrelia, which causes fever, depression, cyanosis and anaemia (spirochaetosis).
- *Ticks*: a heavy infestation can produce severe anaemia and, in extreme cases, death due to blood loss. Argas persicus is particularly dangerous, being the vector of several blood parasites such as the haemoprotozoa and microfilaria. In Malaysia, it was reported that out of 201 blood samples taken from village birds, more than 100 contained Leucocytozoon sabrazesi, 30 had microfilaria, and six carried Plasmodium gallinaceum (Avian malaria). Avian malaria infection is much higher among exotics and cross-breeds.

Internal Parasites (endoparasites)

The more important internal parasites are:

- *Helminths (worms):* these are common in scavenging poultry, especially nematodes and cestodes. Ssenyonga showed that worms were a major cause of lowered egg production of scavenging poultry in Uganda, the most commonly found being Ascaridia galli (Round Worm), Heterakis gallinae (Caecal Worm), Syngamus trachae (Tracheal Worm) and Raillientina spp. (Tape Worm).

— *Protozoa*: the most pathogenic are the coccidiosis disease species of Eimeria tenella and E. necatrix. Coccidiosis is a common parasitic infection in scavenging poultry. It affects mostly young birds, and the most important signs are emaciation, thirst, listlessness, ruffled plumage, bloodstained faeces and birds huddling together. Surveys in Southeast Asia and East Africa showed that 73 and 47 percent of birds, respectively, had positive faecal samples of Eimeria spp. The presence of the coccidia organism in faecal samples indicates an infection, but not necessarily at clinical disease levels. Like antibody presence in blood samples, it may indicate a degree of immunity. This should not be "treated", as doing so eliminates the immunity.

Fungal Diseases

Mycotoxicosis

The fungus Aspergillus flavus commonly grows on stored feed ingredients where moisture content is over eleven percent, especially cereal grains (such as maize [corn]) and oilcake meal (such as groundnut [peanut] meal). The aflatoxin called mycotoxin is the specific toxin produced by A. flavus. The toxin itself may remain after all sign of the fungus mould is gone. Ducks are more vulnerable to the toxin (with a lethal dose in the feed of one part per million [ppm] of aflatoxin) than chicken, which can tolerate up to four ppm. In acute forms of the disease, mortality can be as high as 50 percent. Common adverse effects include immunosuppression, reduced growth in young stock and reduced egg production in hens.

Aspergillosis

This disease is also called airsaculitis. The fungus Aspergillus fumigatus causes the disease by growing as a fungus in the lungs and interconnected air sacs. The fungus grows on damp litter or feed, and the bird breathes in the spores, which grow into easily visible lesions as green and yellow nodules, which can completely fill the lungs.

Non-Infectious Diseases

Deficiencies

Poultry health is also affected by nutritional and environmental factors, such as insufficient feed or feed deficiencies. A high mortality rate among chicks during the first days or weeks after hatching may be caused by insufficient feed and water. A high mortality in adult birds may be due to nutritional problems, such as salt deficiency.

Energy and protein deficiencies and imbalances can arise when the feed contains insufficient quantities of these nutrients, resulting in poor growth in young stock and a drop in egg production and egg weight in laying hens. Mineral and vitamin deficiencies may result in poor growth, low production or death. Vitamin D deficiency causes rickets (bone deformities) in young chicks and, if combined with a calcium deficiency, in chickens of all ages. A lack of manganese results in deformities of the feet of older chickens.

Toxicities

An excess of certain nutrients, especially minerals, can cause abnormalities. An excess of common salt (NaCl), for example, results in deformed eggshells as well as increased water consumption, and if drinking water is restricted (as is often the case with free-ranging birds), signs of toxicity may develop. Free access to feed of high carbohydrate and low fat, combined with lack of exercise, high temperatures and stress, can cause Fatty Liver Syndrome, which can result in high mortality.

Ingestion of toxic plant parts (such as leaves, seeds and sap) is a common hazard for free-range birds. Some toxins are produced by microorganisms, such as those liberated by the bacteria Clostridium botulinum and C. perfringens, both found in the soil. C. perfringens causes necrotic enteritis, caused when the bacteria multiplies in the favourable conditions of the digestive tract and liberates a potent toxin that results in high mortality.

Occasionally affected birds show anorexia, depression and diarrhoea, but most die without showing any clinical signs. C. botulinum causes botulism disease, which is acute food poisoning. This is more common in ducks, which show the nervous symptoms of neck bent down and feathers falling out easily when lightly pulled. Botulism results from the bird eating rotting vegetable scraps, which contain the toxins produced by the C. botulinum. Household vegetable scraps which are not regularly removed are a potential hazard for botulism.

Epidemiology

Management System Effects

Although nearly all the important poultry diseases are found under all types of management, the pattern of disease in free-range birds is different from that seen in intensive poultry production. Free-range flocks usually comprise different species of all ages, and are constantly exposed to the weather, environment and seasonal outbreaks of disease, as well as to germs and parasites found in the soil and in wild birds and animals.

In a 15-year study of the incidence of poultry diseases in northern Nigeria, Sa'idu et al. found viral infections (such as ND in chickens and pox in turkeys) to be the most common cause of disease, although concurrent viral infection with parasites constituted about half the cases studied. They concluded that viruses and parasites caused the most important diseases in indigenous chickens and that they were seasonal in their onset.

Another study, which looked at the changing pattern of diseases in southern Nigeria over the period from 1949 to 1955, found that mortality in the free-ranging flocks at the University of Ibadan was due mostly to the following:

- Helminthiasis, due to Raillietina, Heterakis, Ascaridia, Capillaria, Tetrameres and Syngamus spp.;
- Pediculosis, due to Menopon, Gonoeodes, Goniocotes and Lipeuris spp.; and tropical poultry mites

(Ornithonyssus bursa) in chickens, and Numidilipueria tropicalis in guinea fowls.

These parasitic infections were greater causes of mortality than Newcastle Disease at that time.

Another survey of backyard chicken flocks in Zimbabwe tested 450 blood samples from 52 flocks, and found Infectious Bronchitis (IB) in 85 percent and Newcastle Disease (ND) in only 27 percent of the samples (Table 4). A possible explanation for the lower frequency of ND is that the mortality is usually much higher from ND than for IB, so fewer birds survive to be counted.

Table 4. Diseases affecting local chickens and turkeys in Zaria, northern Nigeria

Chickens	
Disease	Proportionate mortality (%)
Newcastle Disease	36.1
COMBINATIONS	28.5
Snake bite	8.6
Other Diseases	8.6
Gumboro	7.1
Fowl Pox	5.1
Ectoparasites	3.5
Endoparasites	2.5
Turkeys	
Disease	Proportion of all diseases (%)
Turkey pox	16.5
Ectoparasites	15.7
Newcastle Disease	12.2
COMBINATIONS	10.6
Infectious sinusitis	10.2
Endoparasites	3.5
Other Diseases	31.3

Table 5. Disease status in backyard chicken flocks in Zimbabwe as shown by blood testing

Pathogens	Percent positive samples
Infectious Bronchitis	85
Reticulo-endotheliosis	65
Gumboro Disease	55
Pasteurella multocida	52
Mycoplasma gallisepticum and /or Mycoplasma synoviae	33
Newcastle Disease	27
Encephalomyelitis	11
Avian Leucosis	9
Reovirus	3

Similarly, in Zambia, a survey based on 2000 blood samples found that the mean seroprevalence of Newcastle Disease was 37 percent, which varied between 29 percent in the northern province and 51 percent in the Copper-belt province.

A summary of the relative importance of poultry diseases gathered from other sources is tabulated in Table 5.

Poultry Species Effects

In tropical countries, Newcastle Disease is considered to be the most important disease of village flocks because of its high mortality, which is above 70 percent in most African countries. However, not all poultry species are equally susceptible. Guinea fowl, although sometimes affected, appear to have better resistance to ND even when kept with chickens. They are however more susceptible to Trichomonas, to which chickens appear to be immune. Ducks are not thought to be susceptible to ND (although they are significant carriers of the disease), or to most other common diseases of chicken, but they succumb easily to diseases specific to ducks, such as:

— Duck Virus Enteritis (Duck Plague): this is an acute and highly contagious disease of ducks, with a mortality rate of up to 100 percent.

— Duck Virus Hepatitis and Mycotoxicosis: these pose a great danger for ducklings. Mycotoxicosis is mostly contracted from ingesting aflatoxin from mouldy compound feeds; a dose of 0.75 ppm can kill a duckling.

— Duck Cholera (Pasteurella multocida infection): this is a widespread disease of ducklings in village flocks, and chickens can also be infected.

— Pasteurellosis (Pasteurella anapestifer): this is an important disease of ducklings, while Escherichia coli infection (Colibacillosis) is a septicaemic disease of growing ducks. Chickens can also be infected with E. coli.

The above diseases are largely responsible for the huge losses of ducklings that are a feature of free-range duck production.

Although salmonellosis is not a major disease in ducks, duck eggs are a significant source of salmonellosis in humans. This may explain the taboo against touching duck eggs prevalent in many cultures, particularly in Africa. Duck eggs should not be stored in contact with vegetables that will be eaten raw, such as carrots, lettuce and cabbage.

Season and Age Effects

Disease patterns vary according to the season. Newcastle Disease is more serious during the dry season. In Thailand, Pasteurellosis, Coryza and streptococcal infections also occur more frequently in the dry season, and Fowl Cholera, Colibacillosis and Pseudomoniasis in the rainy season. In northern Nigeria, where Sa'idu et al. studied 522 cases involving 8 800 chickens, ND accounted for 30 percent of all cases.

Of these, 38 percent occurred immediately before the dry season of October to December, and only 10 percent during the rainy season of July to September. As for Fowl Pox, more outbreaks occurred in the rainy season and were highest in the month of July, and about 60 percent of the outbreaks affected young chicks. In Thailand, Thitisak et al. (1989) noted that a catastrophic mortality had occurred in March in both 1987 and 1988, this being the late dry season when early storms cause sudden drops in temperature which chill the birds. They also found that chickens under two months of age (normally a rapid growth-rate phase), and those over six months of age (in the process of becoming sexually mature) were more susceptible to infectious diseases.

Epidemiological Studies

Epidemiological studies of village poultry are essential for the development of appropriate village-based poultry health programmes. These have been attempted in many countries, but the work undertaken in Thailand will be used as an example.

The North-Eastern Regional Veterinary Research and Diagnostic Centre, Tha Pra, Khon Kaen, Thailand, with assistance from the German Agency for Technical Cooperation (GTZ) and the Department of Veterinary Clinical Sciences of Massey University, New Zealand, embarked on a long term-study ("Health and Productivity of Native Chickens"). This was part of a programme of epidemiological investigations of factors affecting livestock productivity in the region. The objective of the study was to decide on the priority of the problems for which control programmes could be developed within the regional Basic Poultry Health Service.

In selected villages, about 15 families with flocks of 15 to 20 birds were recruited. In most of the villages, vaccinations against ND, Fowl Pox and Fowl Cholera were carried out in order to encourage interest. Two villages were paid to represent a totally unvaccinated control population. Birds were wing-tagged and grouped by age:

- — under two months;
- — two to six months;
- — six to twelve months;
- — one to two years;
- — two to three years; and
- — over three years of age.

During each visit, the number of eggs and chickens were weighed, counted and scored, and blood was collected from the wing vein for determination of ND titre and for Pullorum, CRD, IB and Gumboro (IBD) tests. Each tagged bird was subjected to health, feather and ectoparasite scoring. Sick and dead birds were collected for pathological examinations at the laboratory, in order to identify the cause of death or illness, and refrigerators were provided in some villages to store dead birds for later examination.

Dead or diseased birds were exchanged for healthy birds in order to examine as many as possible, but as the villagers would sometimes eat dead or dying birds, most of those exchanged tended to be the immature birds which owners were less willing to eat. By calculating age-specific mortality rates and then determining cause-specific rates within each age-group, the contribution of each disease to the mortality within an age-group was determined. A questionnaire was used to gather additional information.

The objectives of the survey were to establish:

— a productivity index;

— population dynamics;

— the importance of common diseases, their incidence and prevalence;

— the average life span of birds;

— patterns of disease outbreaks; and

— the relationship between disease and production levels.

The parameters used for the survey were:

— egg production;

— egg hatching rate;

— number of deaths in each age group;

— weight gain of growers;

— number of birds killed for eating;

— number of birds sold; and

— unexpected losses.

Poor management of eggs, hatching and young chicks, as well as malnutrition, particularly in the dry season, were found to be important predisposing factors to infectious diseases and parasitic infestation.

Disease Control in Family Flocks

Non-medical Disease Control

The most economical and effective means of preventing non-viral diseases is improved management and nutrition, of which the most important aspects are hygiene, housing, flock structure, and young chick care and feeding.

Hygiene

The following simple hygiene measures, which help in disease prevention, were recommended by the FAO/UNDP Small Stock Development Project in North Kivu, Zaire:

- Droppings, feathers and dead birds are sources of pathogens and should be removed from overnight housing and the free-range compound, and then properly disposed of. This will also reduce the incidence of external parasites.
- New arrivals to the flock should be isolated. Birds bought or received as gifts should be quarantined in a basket or cage for at least 15 days; if they remain healthy, they can then join the flock.
- All new arrivals should be treated for ectoparasites and endoparasites as well as vaccinated on arrival if possible.
- Sick birds should be isolated or slaughtered promptly, and dead birds buried.
- The litter in the poultry house should be turned frequently and changed if wet.
- Overnight security baskets should be put in the sun to dry properly or suspended near a fire during the rainy season.
- Feeders and drinkers should be cleaned frequently.
- Broken pots used as drinkers should be heated over a fire before refilling.

- The poultry house or basket should be regularly disinfected every two months.

Improvements in Housing

Simple improvements and maintenance can be carried out when the poultry house is not in use. Important factors in good housing are:

- Ventilation: if poultry baskets are used for overnight housing, they should not be covered with cloths or sacks. Huts, coops and baskets should not be placed near dunghills or pit latrines.
- Proper spacing: overcrowding should be avoided, and numbers of poultry should be restricted to the space available. Weaned chicks and growers should be kept in separate overnight housing. Laying and brooding nests should be left undisturbed.
- Separate species: it is better to keep only one species of poultry but if this is not possible, the species should be housed separately overnight to avoid the spread of disease.

Flock Structure

Of all the common free-range poultry species, chickens are the most susceptible to disease. Ducks, geese and guinea fowl are often symptom-less carriers of chicken diseases, or have mild forms of them. This represents a common source of infection in chickens, while the opposite is rare. Therefore in mixed flocks special attention should be paid to the health of chickens. Separation into different species and age groups may not be possible, but simple devices such as creep cage-baskets may be used as a temporary measure for procedures such as vaccination of chicks or special feeding.

Feeding

The importance of nutrition in flock health is well known. There is a need for further research into alternative feeds for rural poultry, which avoid the use of grain for human consumption.

Medical Disease Control

Simple medical control measures appropriate for free-range village flocks include:

— Vaccination against Newcastle Disease, Fowl Pox and Fowl Cholera.

— Deworming for internal parasites in a mixed flock, with a polyvalent poultry dewormer such as Piperazine (added to drinking water). With guinea fowl, a dewormer against Trichomonas should be used.

— Treatment for external parasites. Insects and other external parasites build up quickly in poultry huts, coops and baskets. There are effective traditional methods against ectoparasites. All the surfaces of the basket, coop or hut can be sprayed with a suitable insecticide, using the same type of hand-pump used for spraying mosquitoes. This procedure should only be carried out when the house is empty in the morning, and the birds should not be allowed back inside until evening. External parasites living on poultry can best be treated by adding powdered mothballs (naphthalene) and ash to the dust bath area. Ash dust is more abrasive than ordinary soil dust, and thus removes the waxy coating of the insect exoskeleton when the bird takes a dust bath. If enough of the waxy coating is removed, the insect will dehydrate and die.

Newcastle Disease Control

There are three general approaches to the control of ND:

— *Hygiene*: this is always important, especially in the control of ND in semi-intensive systems where birds are confined within a fenced yard or house. Hygiene includes measures such as cleaning, disinfection, limiting access to wild birds, and personal hygiene of the farm staff.

— *Slaughter of infected flocks*: this is a drastic measure, which has been successfully employed in isolated regions or islands that are essentially free of the disease.

— *Vaccination in combination with appropriate hygiene measures*: this remains the most effective way of controlling ND.

Newcastle Disease Vaccines and Vaccination Campaigns

For viral diseases, vaccination is the only form of prevention. A proper vaccination campaign can rapidly and significantly minimise losses due to disease. In Indonesia, after an ND vaccination campaign, mortality in village flocks dropped from 50 to 8 percent and the population of chickens increased from 900 to 3 500, representing a 250 percent increase. ND vaccines are available in either "live" or "dead" forms:

— Live vaccines are fragile and have very precise rules for use, requiring a cold chain up to the point of application to the bird. Their effectiveness is reduced if there are residual antibodies in the chickens. This is especially important with maternal antibodies, which are retained by the newborn chick and protect it for up to ten days. Even a low level of maternal antibody reduces the effectiveness of gaining immunity from the vaccine. Group vaccination can be administered in very clean drinking water in very clean drinkers, or by aerosol (in enclosed buildings). The conventional live vaccine, Hitchner B1, cannot be given in drinking water to village flocks, but can be given using the eye-drop method, which has the advantage that each bird receives its dose individually. This has been successfully carried out in Morocco, where it led to a considerable reduction in mortality. The eye-drop method should be used only if there are veterinary personnel available for training vaccinators.

— Killed vaccines give good immunity but require priming with a live vaccine for best results, unless a natural infection has already served this purpose. They have been used successfully in Burkina Faso. Killed vaccines have two disadvantages: they must be administered individually by intramuscular injection, which requires some veterinary training, and—as with live vaccines given by eye-drops—the birds must be caught, a cumbersome task which cannot be avoided with the techniques presently available. Killed vaccines have the advantage that they do not require as rigid a cold chain as do live vaccines, and, because they have a consequently longer shelf life, they can be used in more remote locations. They appear to be most effective in birds that have already acquired some degree of immunity from natural NDV exposure or an initial live vaccine inoculation.

Another advantage of killed vaccine is that the virus-killing chemical used in its preparation also acts against all other possible vaccine pollutants, such as unwanted viruses, bacteria and other microorganisms. Killed vaccines are usually cheaper than live vaccines because the product is more durable, but this is only viable for large flocks. Evidence from Burkina Faso and Niger indicates that because each vial contains at least 100 individual doses, there was a high degree of wastage, as the villagers only managed to vaccinate a few dozen birds a day at best. Much of the advantage gained in efficient manufacturing, packaging and dispatching can be lost at this final stage if the contents of the vial are not fully utilised.

Constraints to Rural Flock Vaccination

The low success rate of ND vaccination is almost entirely due to inactivation of the vaccine because of the absence of an efficient cold chain. This in turn is aggravated by the scattered distribution of village flocks, bad road conditions and lack of transport. In Indonesia, the period between the vaccine leaving a central laboratory and vaccination in the village can be several days.

Vaccination programmes should be carried out at appropriate times. There are seasonal patterns to outbreaks of ND and Fowl Pox, the diseases for which vaccination campaigns are usually carried out. The farming programme should be taken into consideration. In Thailand, for example,

ND vaccinations are carried out in the dry season when the farmers are not involved with rice cultivation.

It has frequently been said that lack of motivation is a major cause of the low vaccination rates in rural areas. To overcome this problem, a pilot project in Thailand organised a training course in primary schools for children of 12 years of age. They were introduced to the concept of the advantages of vaccination against ND and taught to recognise the simple clinical signs of the disease. Another training course for livestock volunteers was given to five selected young village leaders. Upon return to their village, they gave their services and advice on ND control to the village, free of charge. It was hoped that the increased knowledge and commitment would result in better motivation of the villagers to develop their own vaccination programmes.

In Bangladesh, the subdistrict livestock officer organises special training in vaccination and livestock husbandry, in consultation with the local subdistrict committee chairman and members, who then select the farmers and volunteers for vaccinator training. The course is divided into two phases: theoretical and practical. On completion, the vaccinators are supplied (at cost) with vacuum flasks and other necessary equipment, to vaccinate their own and other villagers' flocks, for a fee. They then return the empty vials and receive fresh vaccine. The Department of Livestock Services runs the livestock disease control programme for small farmers as part of its poultry development programme, usually with the assistance of local non-government organisation (NGO) groups.

ND V4 Vaccine

This is a new vaccine, which has the following advantages:

— It is a heat-tolerant vaccine selected from non-virulent forms of NDV naturally occurring in Australasia. In an experiment conducted in Malawi, young chicks which received the vaccine after it had been exposed to ambient temperature for six weeks developed high antibody titres and resisted challenge with a velogenic, viscerotropic ND virus strain.

— After about half of the birds of a household flock are caught and vaccinated, and if all the birds are then confined together overnight, the vaccine will spread naturally from vaccinated to unvaccinated birds. The antibody response of these "naturally" vaccinated is comparable with that of the vaccinated birds.

— Unlike conventional vaccines, which cannot be given if the birds are under stress, the ND V4 is so mild that it can be given to birds under stress.

— The vaccine can be administered by mixing with feed, although the eye-drop method of inoculation is much more effective.

Alders et al. reported a laboratory vaccine trial with the heat tolerant V4 and the Hitchner B1, in which vaccination with the Heat Resistant (HR) V4 gave slightly higher HI (Haemaglutination Inhibition) titres than vaccination with the Hitchner vaccine. The live HRV4 vaccine was used successfully in the control of ND in village chicken flocks in Malawi. In these field trials, vaccinated birds showed a good immune response when the vaccine was administered through very clean drinking water in very clean drinkers. However, this method can only be used in the dry season, because of the difficulty of confining the birds in completely dry conditions during the rainy season.

A Heat Resistant derivative of the V4 vaccine added to feed was used successfully in Southeast Asia. It did not require individual doses, and it spread between birds to some extent. The choice of feed to be used as the vaccine carrier was crucial. Commercial feed as a vaccine carrier has two disadvantages: firstly, its composition varies, and certain components can prove toxic to the vaccine virus; and secondly, prepared feeds are expensive, and one of the essential factors in a smallholder poultry vaccination scheme must be low cost.

In Malaysia, the vaccine is sprayed onto wheat while it is being mixed in a simple, locally manufactured auger-driven mixer. Up to 10,000 doses of vaccine can be mixed at a time. The feed-virus mixture is then put into small plastic bags, each containing 200 g (enough for 20 birds). Even after 45 days storage at ambient temperatures, the bagged vaccine provides up to 90 percent protection.

In other Asian countries where wheat is not readily available, rice has been used. Both rice and wheat have a water-soluble viral inhibitor in the grain coat; thus washed grain maintains a higher virus titre than unwashed grain. Washing is carried out by soaking the grain for 24 hours and then mixing the vaccine with the wet grain. At the village level, the required amount of washed grain (10 g per bird) is measured into a plastic bag and the vaccine added. After mixing thoroughly, the feed-vaccine mixture is fed directly to chickens. It is important to provide creep-feeding for the chicks and growers, otherwise the older birds get all the vaccine.

The V4 strain is being tested in several African countries (Ethiopia, the Gambia, the United Republic of Tanzania, Zimbabwe and Nigeria). The first field test of HRV4 (Heat Resistant V4) was conducted in the Gambia. In July 1993, FAO and the governments of the Gambia and Ethiopia signed a Technical Cooperation Programme project agreement, "Assistance to Rural Women in protecting their chicken flocks from Newcastle disease" (TCP/RAF/2376T).

The objectives were to introduce and evaluate, under African rural conditions, the HRV4 oral vaccine, which had proved effective in Southeast Asia, and to involve rural women in the implementation of the project. In both countries, repeated oral vaccination (up to four times at regular intervals) did not generally produce the high immunity found in Southeast Asia. However, parallel groups inoculated by eye-drop did achieve high levels of immunity.

In 1995, confined laboratory trials of ND HRV4 feed-based vaccines were carried out in Ethiopia. Oral vaccination, administered in barley, wheat and crushed maize feeds, was compared with eye-drop vaccination and with an unvaccinated control.

Eight days after the birds were challenged by the ND virus, the survival rates were: 80 percent of the eye-drop vaccination group; 20 percent of the oral vaccination groups; and none of the control group. Of the three feeds used in the oral vaccination trials, barley proved to be the most efficient vaccine carrier in terms of providing disease resistance, followed by maize and lastly wheat.

Zimbabwe was free of ND from 1986 until 1994, when suddenly infection was introduced from South Africa. Because of unrestricted poultry transport, it spread rapidly across most of the country. During 1994/95, ten million rural chickens nationwide were vaccinated at a project cost of US$1.50 per bird. In early 1996, a FAO-assisted Technical Cooperation Programme was initiated to establish a community-based programme for the prevention of ND epidemics in rural chickens.

The approach combined the immediate administration of the V4 vaccine using the conventional eye-drop vaccination method with community-based trials using the feed-mixed oral method under Zimbabwe conditions. Thus it was hoped to avert other epidemics and generate data for planning regular feed-based vaccination of rural chickens as part of the veterinary extension service.

Although V4 has shown promising results in overcoming the major constraints associated with ND vaccination, there are still problems with its use in scavenging systems at the village level, where feeding is irregular and poultry are hatched, bought and sold throughout the year. In all countries, laboratory trials to assess various feeds as carriers and to familiarize technicians with the vaccine and virulent challenge systems should be conducted before field trials. Production data should be collected before, during and after vaccination. This is essential to evaluate the efficiency of vaccination.

Traditional Poultry Disease Control

Traditional treatment and control of disease is important, as most developing countries cannot afford to import or subsidize veterinary drugs and vaccines for smallholder farms. There is also increasing concern about the effect of synthetic drugs on animals and the environment. Ojeniyi found a correlation between the use of antibiotics and drug resistance among E. coli strains isolated from intensively managed poultry at the University of Ibadan, Nigeria. All 1248 E. coli strains from the University poultry farm, and 2196 strains from a commercial poultry farm in Ibadan, were resistant to tetracycline, streptomycin and sulphonamide. In contrast, all 2284 strains isolated from free-range town and village poultry were sensitive to these drugs.

Most of the information presented below on traditional medicine used for poultry has been collected informally, and has not been scientifically tested. The main reason for its inclusion here is to encourage formal research.

Viral Diseases

- *Newcastle Disease*: in Nigeria, either Lageneria vulgaris or the bark of Parkia filicoidea are given to the flock in drinking water. In Zimbabwe, the leaves of Cassia didymobotrya or the latex of Euphorbia matabelensis are given in drinking water. In the United Republic of Tanzania, in the regions of Arusha and Kilimanjaro, the stem of Euphorbia candelabrum Kotschy var. candelabrum or the fruit of Capsicum annuum together with the leaves of Iboza multiflora are given.
- *Fowl pox*: in Zimbabwe, the leaves of Aloe excelsa are soaked and the extracted fluid is added to drinking water. A poultry disease named "Yoko yoko" by the Fulani of Mauritania, Mali and Senegal, is a serious epidemic, affecting hens of all ages. The exact cause of the disease is unknown, but Ba (1982) suggested that it might be a type of Fowl Pox. The signs and symptoms, as described by the Fulani, are dejection, breathing difficulties with the emission of the sound 'yok yok' and sneezing. Obstruction of the nostrils by yellowish crusts makes the birds breathe through their beaks, which are also encrusted. Lack of appetite and purulent conjunctivitis have also been noted. Eventually, the

birds suffocate and die. The mortality rate can be as high as 100 percent in growing chickens, but some adults survive. The birds are systematically slaughtered and the hen house burnt to prevent spread. The symptoms, as described above, are indicative of the "wet" form of Fowl Pox, where eventually the bird suffocates from a cheesy growth in the trachea. The pox lesions are sometimes less obvious in this type of Fowl Pox disease, but they can be found on close examination.

— *Colds*: in Nigeria, Hibiscus subdariffia is pounded, mixed with drinking water and given to birds with ruffled feathers.

Protozoan Diseases

Coccidiosis: in Nigeria, Lageneria vulgaris is dipped in the flocks' drinking water.

Bacterial Diseases

Fowl Cholera: in Nigeria, the fruit of Adansonia digitata is broken and soaked in the birds' drinking water. The fruit of Capsicum annuum is mixed with soot from the ceilings of thatched buildings (Hausa: Kunkunniya) and given in drinking water.

Metabolic and Infectious Diseases

— *Abdominal disorders*: in Nigeria, the young leaves of Boswellia dalzelii are chopped and soaked in water, and the extracted fluid is given as a diarrhoea treatment. In South Africa (Natal Province), Leonotis leonurus Ait.f. is given to treat both yellow and green diarrhoea. In Southern Africa, farmers use a cold infusion of the leaves of Aloe saponaria Haw. to treat enteritis and indigestion in poultry. In West Africa, the chopped leaves of Pergularia extensa are fed to turkeys suffering from diarrhoea.

In Zimbabwe, the bulb of Adenium multiflora is soaked in water for 12 hours, after which the sick animals are drenched to treat for bloody and watery diarrhoea. For the same purpose, the latex of Aloe chabandii or Euphorbia matabelensis is given in drinking water.

— *Blood in the excreta*: in Zimbabwe, the bark of Cussonia arborea is soaked in water, and the sick birds are drenched with the fluid.

Ectoparasites

— *Various ectoparasites and parasitic diseases*: in Nigeria, the dried leaves and twigs of Guiera senegalensis Lam. are burned in poultry houses to reduce ectoparasites. In Bulawayo, Zimbabwe, Thamnosma africana Engl. is placed in chicken pens to repel fleas and ants.

— *Lice*: in Nigeria, the leaves of Bandeiraea simplicifolia are placed in hen houses to kill lice. Ash from the burnt leaves of Nicotiana rustica, N. tabacum or Carica papaya is rubbed into feathers to protect against infestation. In Senegal, the leaves of Calotropis procera Ait.f. are used to kill lice on poultry.

Endoparasites

Worms: in Nigeria, the fruits of Cucumis prophetarum and Solanum nodiflorum are used by the Hausa people to treat poultry for worms.

References

Adene, D. F., "Pre-epidemic and epizootiologic features of avian influenza in Nigeria", *Western Poultry Disease Conference: Proceedings of the Western Poultry Disease Conference*, 2007.

Andreasen, C. B., "Training resources for prevention and intervention strategies during avian influenza", *Western Poultry Disease Conference: Proceedings of the Western Poultry Disease Conference*, 2007.

Lee, C., "Development of live attenuated vaccine against avian influenza", *Western Poultry Disease Conference: Proceedings of the Western Poultry Disease Conference*, 2007.

Weber, T. P. and Stilianakis, N. I., "Ecologic immunology of avian influenza (H5N1) in migratory birds", *Emerging Infectious Diseases*, 2007.

13

Poultry Breeding Strategies

Strategies to develop poultry breeds suitable for family poultry smallholders in tropical countries must differ from those used in intensive production, and should focus on improving indigenous breeds while also making use of pure exotic and crossbred chickens where appropriate.

Conservation of local breeds possessing genetic variations specific to the particular environment is essential for sustainable development. Although they exist as numerically small populations, local breeds are not only highly adapted to the natural environment, but are also an integral part of the lifestyle of the rural people. People, livestock and environment form a delicately balanced but sustainable ecosystem, and thus the potential impact of any intervention to improve production in the traditional system should be predetermined.

The situation is less sensitive in peri-urban, industrial and small-scale intensive poultry production, in which rapid improvements can be achieved through well-designed development programmes. The intensive poultry production sector, however, is generally much smaller than the family poultry sector in virtually all developing countries.

Breed Improvement Strategies

The following two rules should be incorporated into breeding strategies:

— Germplasm in traditional conditions should not be modified until management and housing have been improved and, even then, selection should be restricted to local breeds.

— When technical conditions are optimum and a ready market exists for the products, then improved breeds, crosses and hybrid strains that have been selected for high performance can be introduced into the peri-urban system, even at small-scale levels.

The most common method of improving the local gene pool is crossing indigenous and exotic birds, and then leaving the hybrid offspring to natural selection. Pure-bred or hybrid cockerels (or pullets) selected for greater meat or egg production are introduced into local flocks, usually in order to increase egg production. It is important to note that improved growth (for meat production) and high egg production are genetically incompatible in the same bird. The genetic traits are negatively correlated, which means that selection for one trait will reduce the other.

Cockerel or Pullet Exchange

An example of this type of strategy is a flock of indigenous local hens laying 50 eggs a year and beginning to lay at 25 weeks of age, crossed with "improved breed" cockerels, which have a genetic breed potential of 250 eggs a year, with hens beginning to lay at 21 weeks. The results are crossbred hybrid pullets beginning to lay at 24 weeks, with a genetic potential of laying 200 eggs per year. The first generation hybrid cross-breed has a higher theoretical genetic potential (genotype) than the average (150) of the two parent breeds, due to the effect of hybrid vigour. However, unless management (especially in the area of nutrition) is improved, this genetic potential will not be realised by the hybrid cross-breed in actual practice in the environment.

If subsequent generations of the hybrid crossbred pullets are mated (back-crossed) again with the same "improved breed" cockerel, the genetic potential for increased production is raised, although at a slower rate (as hybrid vigour only works with first-time crossing). With each generation, higher levels of management (including the provision of properly balanced feeds) are required to achieve this potential.

If the hybrid cross-breeds mate among themselves, however, potential production falls in the very next generation to the average potential of the two original genotypes, even if management could support the higher hybrid level.

The use of cockerels in this way is the basis for the Cock or Cockerel Exchange Programme (CEP) or Opération Coq, which has been implemented in almost all tropical countries. Households exchange all their local cockerels for a few improved cockerels, which are then raised to maturity to allow them to adapt to local conditions.

In some cases, a Pullet Exchange or Hybrid Hatchable Eggs Programme is used. These approaches were used extensively from the early 1930s until the 1960s, by which time urban development had begun to give rise to peri-urban, intensive, small- to medium-scale poultry production, which makes use of imported commercial breeds and technology.

The gradual replacement of local genes through crossbreeding and artificial selection has been the basis of initial development in many countries. Although many strategies deemed appropriate for smallholder poultry production systems have been implemented, most have not succeeded, due to a lack of management input to support the improved potential.

Replacement of Indigenous Breeds

The use of hybrid chickens under free-range rural conditions has often been studied, notably in Zimbabwe, and in Sri Lanka, Zambia and Nicaragua.

It has consistently been found that entire flock replacement programmes lead to increased egg and meat production, but only where management supplies good nutrition and veterinary hygiene. There is, however, one great disadvantage, in that the use of commercial hybrids to increase egg production necessarily eliminates broodiness of hens, due to the negative genetic correlation between these two factors. For this reason, complete replacement of local birds should not be considered unless a reliable local supply of day-old chicks (of an appropriate breed) is available.

Selection within Local Breeds

Production Traits of Local Breeds

The genetic development of local breeds and varieties in developing countries first requires proper documentation of their productive and reproductive performance. The main production characteristics of local breeds are:

- small body size (low nutritional maintenance requirement);
- lateness in maturing (up to 36 weeks of age);
- low performance in egg numbers (20 to 50) and egg size (25 to 45 g);
- small clutch sizes (two to ten eggs); and
- long pauses between laying of clutches and a predominant inclination to broodiness.

For rural smallholder extensive systems, meat production cannot be separated from egg or chick production, and thus a highly broody (with consequent low egg production), low body-weight (low-feed requirement) bird is best for survival under these conditions. Surplus cockerels, whatever they weigh, are usually sold for meat when they reach sexual maturity at three to four months of age. Under rural smallholder extensive systems, there is little reproductive control of the hens, as they brood their own chicks for continuous regeneration of the flock. The egg brooding (incubation) and chick rearing activity increases the reproductive cycle length by 58 days to about 74 days in total:

16 days for egg laying and clutch formation + 21 days for hatching + 37 days (5.3 weeks) for chick rearing = 74 days

Thus, most hens can produce chicks about four to five times per year, and only four times if the rearing period is extended to eight weeks. As malnutrition, infections, predators and accidents result in mortality rates of 60 to 70 percent during rearing, virtually all eggs are used for reproduction. With four to five reproductive cycles per year, only about nine replacement pullets may be obtained.

Fertility and hatchability are also high in local birds. They generally adapt well to unfavourable management conditions, and resistance to prevailing diseases is usually assumed to be high, although juvenile and sometimes adult mortality rates can be high in extensive production systems.

Considerable genetic differences exist between different regional and continental populations of indigenous chickens, and production rates of local populations should be evaluated before introducing development programmes.

Selection Programmes for Local Breeds

Although better management procedures can significantly improve the performance of local birds, some feel there is also a need for genetic selection. Pure-breeding and selection programmes have been developed in Bangladesh, although not implemented in the field. Both of the above groups concluded that although improvement of local poultry breeds would be beneficial, it is essential to evaluate breeds and their crosses before undertaking a breeding strategy.

Research conducted in the United Republic of Tanzania concluded that selection for dual-purpose characteristics within individual local populations is both time-consuming and costly. Crossbreeding with improved breeds is recommended, followed by selection in the composite population.

Although consumer preference in most developing countries is for dual-purpose breeds, it is important to restate that in the same bird, the traits of increased egg production and increased broodiness are genetically incompatible, as are the traits of high egg production and high meat production.

Selection for any trait within these pairs will reduce the other trait of that pair.

Modifying Local Breeds

The use of single or combined dominant genes for feather restriction (Na) and feathering structure (F), as well as the sex-linked recessive gene for reduced body size (dw), has been found to be particularly relevant for the tropics. Research into the effects of these genes on economic factors has been undertaken in Malaysia. For example, the feather restriction (Na) or Naked Neck gene results in 40 percent less feather coverage overall, with the lower neck appearing almost "naked".

This considerably reduces the need for dietary nutrition to supply protein input for feather production, and protein is a limiting factor in many scavenger feed resource bases. Barua et al., has reviewed the available information on the performance of indigenous Naked Neck fowl in the hope that it will draw the attention of scientists worldwide to its interesting characteristics and facilitate future research.

The incorporation of such genes could be significant in the development of appropriate breeds and strains for smallholder poultry production in the tropics. There are now seven potentially useful major genes:

— Na - naked neck (autosomal -A);

— Dw - dwarf (sex-linked -S);

— K - slow feathering (S);

— Fa - Fayoumi (A);

— F - frizzle (A);

— H - silky (A); and

— Fm - fibro-melanosis (A).

The use of major genes to improve productivity in smallholder poultry breeding programmes has been researched in various tropical countries (including Indonesia, Malaysia, Thailand, Bangladesh, Bolivia, India, Cameroon and Nigeria).

Other morphological traits that allow better heat dissipation include large combs, large wattles and long legs. Gene coding for these traits, which are not major genes but the result of multiple genes and their interactions, could also be considered for incorporation into the development of high performance local birds for the tropics.

For Better Production

Family poultry is well integrated into most village farming systems, with local breeds representing 40 to 70 percent of the national meat and egg supply in most tropical countries. Because of their scavenger adaptability, production ability and low cost, local breeds are kept by rural smallholders, landless farmers and industrial labourers. It is difficult to imagine birds better adapted for survival under scavenger free-range conditions than the breeds that have already evolved under those very same conditions, and are still surviving as proof of their ability to do so. However, there does remain a considerable and largely unexploited potential for increased production from local breeds through improved management.

The critical management objective for scavenger free-range systems is to reduce the high mortality in both growing and adult age groups, but especially the 60 to 70 percent mortality in the growers. This high mortality means that many eggs laid by the hen need to be used for reproduction to maintain flock size, instead of for sale or consumption. It also means that many birds that die could instead be sold or consumed as meat.

Breed improvement to increase meat or egg production would not solve the health and nutrition management problems. However, increased egg production (by breed improvement) would create a new problem - lack of broodiness in the flock - which would force the smallholder to buy stock rather than have the hen brood and rear her own.

Mortality can be significantly reduced through increasing farmer awareness of health needs, through the provision of vaccine and through improving the nutrition of growing stock. These are the most important improvements to management activities that will enable to the farmer to best exploit the existing potential of local breeds under scavenging free-range conditions.

If management resources available to the smallholder or landless farmer increase to the extent of a local supply of balanced poultry feed, the options open to his income-generating ability are increased. The performance of local breeds will increase slightly under cage or deep litter management but, because the genetic potential for egg production (or meat production) of local breeds is lower than that of commercial hybrids, the same investment in intensive management will achieve a much higher production result by using commercial hybrids.

If balanced feed, good health-care supplies and day-old chicks of hybrid varieties are locally available, then intensive poultry management is an option. If these are not available, raising local breeds under scavenger free-range systems is still the best choice. The vast potential for increasing income generation from scavenger free-range family poultry clearly lies in the management area of reducing mortality in growing chickens. This alone is sufficient challenge for the already overstretched resources of government and NGO field extension staff in developing countries.

The potential for breed improvement is a factor to be considered in the future, but only when the more immediate objective of reducing mortality is attained. Meanwhile efforts should be continued to preserve germplasm as a resource for the future.

References

Hays, F. A., *Poultry breeding applied,* Poultry-Dairy Publishing Co., Mount Morris, Ill., 1888.

Klober, K., "Sustainable poultry for pasture", *Small Farm Today;* 15(2):21-22, Apr/May 1998.

Thear, K., *Free-Range Poultry,* Farming Press, 1997.

Traupman, M., "Profitable poultry on pasture", *The New Farm,* May/June 1990.

14

Marketing of Poultry Products

As a country develops, more of its consuming population lose touch with the village and food producers. Thus more specialised marketing services are needed. Farm produce must be collected, packed and transported in good condition to the cities and distributed to retailers near consumers' homes. This also calls for grading and storage of the product. The more developed the country becomes; the greater is the variety of products that can be economically produced. All this must be provided at a cost that consumers can afford.

A study of existing marketing systems in a country will often reveal how they have evolved to their present state. Many developing countries do not have refrigeration as a factor in their storage, either during transport, retail or consumer household stages. For this reason, poultry meat is purchased live, and slaughtered immediately before consumption. Also, eggs are often retailed with a means for the buyer to check their quality before buying, either by "candling" or a bucket of water (to test the egg's age by the floatation method). Both methods essentially test for the size of the air-cell situated at the blunt end of the egg, which increases in size as moisture is lost from the egg. With a bigger air-cell, there is more floatation.

In developing countries, transport of eggs and poultry from the village to the city usually begins with a purchase by a middleman dealer, direct from the household, or from small locally held weekly markets. Baskets with layers of straw protect the eggs from breakage, and other types of baskets are used to carry live birds. Bullock carts are still used in many countries for transport of both live poultry and eggs to larger community centres. The roofs of buses or trains replace these slower vehicles as transport systems develop. Marketing quality considerations for live birds are usually concerned with weight loss in the bird from dehydration during transport. These are easily resolved by providing drinking water during the trip, and travelling during the cool part of the day when possible.

Improved marketing programmes must add no more cost to the product than the consumer can afford. Important marketing improvements can often be simply made by making small corrections to already existing handling, transport, packaging, grading and storage methods.

Marketing organisations generally come into being very gradually, and must be appropriate for the background, character and education of the people concerned. Plans for radical changes, which do not take sufficient account of social and economic environments, are likely to fail. Thus any improvement programme should be designed to achieve desirable modifications in existing commercial facilities (and their economic and legal framework) by a process of steady growth.

As a country develops, the task of marketing eggs and poultry will still involve the collection of live poultry and eggs from farmers, transporting them to a grading, packing or processing plant, grading and standardising the poultry meat and eggs, processing them and packaging them into more useful forms, storing them (preferably under refrigeration), moving them through wholesale and retail channels and delivering them to consumers at a convenient time and place.

Factors Affecting Demend for Poultry Meat and Eggs

Ceremonial and Traditional Aspects

In traditional societies, poultry are often used for ceremonies, sacrifices and gifts. What follows are some traditional aspects of poultry keeping from the Mossi of Burkina Faso (West Africa), the Mamprusi of northern Ghana, and Bangladeshi and Malay farmers in South Asia.

Among the Mossi people when no poultry is available (such as after a Newcastle Disease outbreak), to meet customary family obligations, the household must purchase or borrow a bird. Chickens are given to convey value to a relationship, or to offer thanks for a favour or help (such as from government officials). For most sociocultural and religious purposes, the required sex and colour of fowls are also prescribed. For example, a family will give a white cockerel when an agreement for marriage is reached.

The consumption of eggs in Mossi villages is uncommon. There is a strong belief that a child who regularly eats eggs will become a thief, reasoning that the good taste of eggs will make the child want to eat eggs often. The only eggs consumed are those that fail to hatch under broody hens. These are boiled and then eaten. Chicken eggs, unlike guinea fowl eggs, are not part of the trade in poultry products, since all eggs are required for hatching to maintain the flock (given the normally high losses during rearing). Dealers from urban areas reflect the demand for village eggs. The eggs are often bought by small food stall merchants who boil the eggs and resell them as snack food. A considerable number of guinea fowl eggs are collected by the Mossi for sale, most of which find their way to the cities via village markets, where dealers buy the eggs.

The Mamprusi society in northern Ghana has a variety of uses for poultry products. Chicken cocks are the most popular sacrificial animals. Guinea fowl cocks are not used. The colour of the bird is important. A red cock is sacrificed to ask for rain or a good harvest; a white cock is used to convey value in relationships, and a black cock is used to ask for protection against disease, war or quarrels. Because of these customs, red, white and black cocks have double the value of cocks of other colours.

The sale of young birds and eggs takes place in the Mamprusi village markets. Prices fluctuate during the year, and are low during the pre-harvest season, when the granaries are empty and the crops are still growing and thus cash is less available. At such times, traders from the south come to buy for resale in the cities. Sometimes, middlemen dealers are involved. They buy the birds in the villages and sell them at markets or to city-based traders. The sale of poultry products from Mamprusi households contributes about 15 percent to Mamprusi annual cash income.

Poultry consumption by the household is rare, as most birds are sold for income generation. In Mamprusi society, women, circumcised girls and first-born children do not consume eggs or meat. These products are only eaten by elderly men, male visitors and young children. The reasons are not fully understood. Some Mamprusi women believe that during pregnancy, their behaviour (including their food choices) can affect their unborn child.

In Bangladesh, eggs and meat are consumed mainly by men and boys, and very rarely by women and girls. Low-income groups generally do not consume eggs or meat. These products are sold, and from the proceeds, essential items are purchased, such as carbohydrate and low-cost vegetable protein foods.

Guinea fowl, more than chicken, are given as gifts to visitors. To give a gift is considered to be a wealth-increasing action as well as an act that conveys value on the receiver. Farmers often save for agricultural equipment or other materials and small livestock is used as a savings account. The offspring, like chicks, are considered to be the interest on the savings.

In many parts of Africa, birds are sold to meet unforeseen expenses, for example, to buy the beer and kola-nuts customarily given to gravediggers when a family member dies. The birds usually sold from the village flock are: surplus males (cockerels and cocks); pullets; old hens; non-productive hens; large-sized birds and sick birds. Young birds are often sold just before the onset of the high-risk period for Newcastle Disease.

Traditional Taste Values Placed on Poultry Meat

It is important to understand traditional taste values and their effect on market demand. The market price for free-range birds for meat is usually stable because:

— the meat is considered tastier and stronger flavoured than commercial broiler meat;

— the meat (muscle tissue) is tougher, and retains its texture when prepared in dishes requiring longer cooking; and

— the birds are not fed with compounded feed which may contain antibiotics, anti-mould compounds, enzymes, sulpha drugs and other medicines or synthetic chemicals.

In eastern Asia, it is believed that chickens fed with chemicals and drugs have poorer therapeutic value, as they do not combine well with ginseng and other oriental herbs used in making soups, especially steamed types. For this type of soup, younger pullets are preferred and thus they fetch a higher price than do the cockerels. The female is said to be more beneficial and the meat tastier. Steamed chicken soup is believed to provide virility and vigour. It is commonly recommended in Malaysia for pregnant women and for those recovering from sickness.

In the case of large-scale commercial ayam kampung (local village chicken) production in Malaysia, local birds are confined and fed on commercial rations but they fetch lower prices than free-range local birds. Such large-scale production has an affect on the market value of all local birds, as purchasers have difficulty distinguishing between genuine free-range and commercially fed local birds. However, the price of ayam kampung continues to hold a margin above that of commercial meat chickens. The introduction of more appropriate methods of Newcastle Disease vaccination in Malaysia will reduce mortality at the village level which may also stimulate further interest in family poultry production. If this happens, there will be an increase in the supply of local free-range poultry products to the market, and the price of the ayam kampung product (from large-scale commercial production) may fall further.

Carcass Parts and Organ Meats

The value of birds for sale in developing countries depends firstly on the available supply, secondly on the age and sex of the birds, and thirdly on their size or weight. Young birds, especially cockerels up to six months of age (weighing up to one kilogram live weight), are usually preferred by consumers. This is because larger birds are more expensive for most households, and smaller birds are more tender and have the same preferred portions (drumsticks for example). Table 1 shows carcass characteristics of the local village chicken in Bangladesh.

Table 1. Product characteristics of indigenous scavenging chickens in Bangladesh

Characteristic	Mean
Live weight, kg	1.14
Carcass weight, %	55
Eggs/hen/year	35 - 45
Egg weight, g	35 - 39
Hatchability of eggs, %	84 - 87

Whatever the size of bird, all chickens have an equal number of high-demand portions (such as breasts and drumsticks), and a similar proportion of gizzards and other desirable organ parts. Buying small birds supplies the same number of the desirable parts for a lower price. Together with the tenderness of the meat, this explains the heavier trade in young birds, which are also bought for replacement stock in depleted flocks.

Supply Mechanisms for Poultry Meat and Eggs

Depending on the location of the farm dwelling, birds and

eggs are sold from the household to traders (dealers or middlemen), direct to consumers, or carried by the farmer to the local market. The role of traders in the marketing of poultry products is an important one. Traders from urban areas buy eggs in villages to sell in cities. Where transport is an important consideration (as in many parts of Africa), guinea fowl eggs, with their stronger shells, are preferred to chicken eggs. Prices of eggs are related to supply and demand, to the higher risk of spoilage and lower use for hatching in hot and humid seasons, and to the availability of alternative protein foods such as fish. There is a tendency to hatch less in the hot season, due to low hatchability and diseases of young chickens, and there is also less hatching in the cold season, due to the risk of chilling stress to the young chicks.

Birds are either brought to the local market once or twice a week for sale to local consumers, to other local markets, or to local traders. Chickens are transported to the market in open-weave (well ventilated) baskets or wooden crates. They need not be fed on the day of sale, but should receive drinking water. If the trip to the market takes eight hours or more, stops should be made to supply water to the birds. In hot seasons, it is better to transport birds at night or in the cooler early morning. While the price of live birds depends on their size, the price of eggs depends more on number.

It is often assumed that for poultry and eggs, producers get 60 to 65 percent of the market price but this has been found to be false in Bangladesh, where they receive less than this. The role of traders or hawkers is very important, as it makes selling from the house possible, but these traders take up to 35 percent of the market value, with a consequent lower profit for the farmers who are responsible for production. This loss of income has stimulated farmers in many places to organise sales through their own marketing groups or formal cooperatives.

Supply Channels

A study by Adeyanju et al. of the marketing of poultry products in Ondo State (in south-western Nigeria) revealed a large number of transactions and participants. The local channel begins with the producer selling poultry products to retailers who serve the needs of local consumers. In most areas, local consumers also buy directly from producers. The other marketing channel involves wholesalers. They buy poultry products directly from producers and sell to retailers inside and outside the State, and are based in urban centres where urban-based consumers are located.

Marketing channels for family poultry often cross international boundaries and can generate significant foreign exchange for the producing countries.

Planning

Forming a marketing plan means identifying where and when birds and eggs will be sold to receive the best possible prices. Putting large numbers of birds up for sale in a small community may depress the price. Even the sale of small numbers of intensively managed layers needs advance planning. A flock of 20 hens may produce 1 200 eggs in a year, even at the low production rate of 35 percent. The plans of other farmers must also be considered. If they all expand their flocks and have good years, prices will almost inevitably fall. Seasonal considerations enter into market plans as well. In India for instance, eggs are thought of as a heat-producing food and are eaten in the cool, rainy season. Many factors affect the quality of eggs and hence the price that consumers are willing to pay for them in the market.

Egg Quality Considerations

Quality determines the acceptability of a product to potential purchasers. The quality of eggs and the preservation of this quality during storage is a function of their physical structure and chemical composition. A basic outline of the most important factors of concern in egg quality is presented below.

Egg Composition

The egg consists of shell, two shell membranes, the white (or albumin) and the yolk. The shell is quite porous to air and water vapour but is very resistant to invasion by microorganisms as long as it is clean and dry. A thin outer covering on the shell called the "bloom" or "cuticle" (which is unfortunately easily removed by washing), assists this process. After the egg is laid, its contents shrink, both from cooling and water evaporation. Air is drawn in (along with anything else on the shell, such as bacteria or fungi) through the pores in the shell to replace this loss.

A gap opens up between the two membranes because the outer one is attached to the shell and the inner one is attached to the egg white. This gap is known as the "air cell" and is usually found at the large blunt end of the egg. The egg white takes the form of a "thick" albumin sack enclosing the yolk, with a more fluid "thin" albumin between this sack and the yolk to the inside, and again between the sack and the shell to the outside. These layers provide a barrier to prevent the yolk touching the shell and to provide food for the embryo. Egg white has specific antibiotic effects, which further protect the yolk. Egg white also contains two fibrous cords (the chalaza), which are attached to the yolk and to either end of the egg, which help hold the yolk in the centre and assist in preventing the yolk from touching the shell.

The weight of an egg laid by a local village breed of hen is about 35 g. Commercial hybrids lay eggs of about 58 g weight. The shell comprises approximately 11 percent of the weight of an egg, the remainder being the edible portion. By weight of edible portion, the yolk is 36 percent and the white is 64 percent.

Shell Quality

Eggs of unusual shape are more likely to be damaged during the marketing process, and consumers do not like them. Small thin cracks in the shell, which do not leak, are called "checks". These are usually detected by candling. "Checked" eggs should be sold for immediate consumption, as their storage life is limited. The household usually consumes eggs with leaking cracks, where the eggshell membranes are broken as well as the shell. Brittle, thin-shelled eggs (shells less than 0.35

mm thick) are also unsuitable for transport to market. Dirty eggs must be cleaned by dry or wet methods, and thus have a higher marketing risk because of the removal of the cuticle.

Shell colour is not a guide to egg quality, but there is usually a consumer bias to either white or brown, which must be considered in marketing.

Egg Yolk and Egg White Quality

Consumers prefer the odour and flavour of normal fresh eggs. The yolk should be round, firm and yellow in colour. Local yolk colour preferences may vary and can be easily adjusted by raising or lowering the amount of green leaf material included in the poultry ration or supplement. Egg white normally has a slightly yellow-green tinge and the thick white is slightly cloudy. Consumers are usually critical of blood or meat spots, which can vary in colour from red to grey, and in size from small specks up to one square centimetre. Blood spots are caused by slight bleeding at the time of release of the ovule (yolk) from the ovary of the hen. They may be found in the white or adhering to the yolk.

Deterioration

The interior quality of eggs deteriorates after laying at a rate depending on time and conditions of storage, such as temperature, relative humidity (RH), and the presence of strong smelling substances or other food items in the storage place. Eggs stored at 27 to 29 °C for 7 to 10 days will show deterioration changes similar to the same eggs stored at minus 1 oC and 85 percent RH for several months. The changes are due to water loss, carbon dioxide (CO_2) and the absorption of volatile odours from the environment.

Moisture Loss

Since an egg contains about 74 percent water and the shell is porous, eggs readily lose moisture. A weight loss of 2 to 3 percent is common in marketing and is seldom noticed by the consumer. When losses exceed this level, the air cell is noticeably enlarged by shrinkage in the contents of the egg.

This loss is reduced if the storage humidity is high and the temperature is reduced. Coating the eggs with oil and other substances can also reduce the loss. The ideal conditions for egg storage are about minus 1 °C and between 80 to 85 percent RH. At storage temperatures of 10 °C and above, the optimum RH is 80 percent.

There is a risk of mould spoilage when the RH is too high. Paper pulp egg trays or other packing materials that readily absorb moisture will accelerate moisture losses from eggs. A temperature as low as 10 °C is unlikely to be practical in rural areas of many developing countries. Temperatures between 10 and 15 °C are more practical, but even then, care should be taken when moving the eggs from cool storage into the outside air with its higher temperature, which often causes condensation to form on the shell, with consequent risks of mould and "rot" growth.

Microbiological Spoilage

The contents of the egg are usually sterile when the egg is laid. The main cause of contamination is the washing of eggs. Wetting the shell allows microorganisms on the shell to penetrate and multiply inside. Common indications are green, black and red "rots", mustiness and sourness. The bacteria causing these effects cannot penetrate the shell if it is kept dry. If eggs do become wet through condensation, for example after removal from a cool store into a warmer room, bacteria may then be able to penetrate the shell.

Tainting

Eggs, especially yolks, are easily tainted by strong odours, from such sources as disinfectants, soaps, diesel, kerosene, petrol, paint, varnish and wood preservatives. Other foods, such as onions and citrus products, can taint eggs after only a few days of exposure.

Quality Control and Maintenance

Maintenance of egg quality is a major problem for those involved in egg marketing. Eggs soiled by droppings or the contents of leaking or broken eggs spoil faster than clean eggs. Only good quality eggs should be sent to the market. The simplest way of sorting is to divide the eggs into three categories: cracked, dirty and clean. The cracked eggs should be eaten or sold locally for immediate consumption. The dirty ones should be cleaned and sold locally for consumption within a few days, while the clean eggs can be sent to the major marketing outlet. In some areas, eggs of certain colour or sizes are preferred, and the eggs should be sorted for these qualities.

Production Factors Affecting Egg Quality

The main production factors affecting egg quality are:

- breed and age of the flock;
- type of feed;
- incidence of disease;
- management control of the laying flock; and
- management control of the handling of eggs.

Breed and Age of the Laying Flock

The effect of breed on the egg is inherent in many aspects, including the colour, thickness and texture of the shell, the incidence of blood spots, and the amount of thick albumin. While commercial breeders pay constant attention to these factors, there is little that farmers can do to control them.

After the first season of egg production, hens produce eggs of poorer shell quality and poorer egg white thickness, even though the eggs are larger in size. The rate of egg production is also lower. For these reasons as well as the high meat value of the carcass of the older hen in most developing countries, it is advisable to replace the hens after 12 to 18 months of lay.

Type of Feed

A balanced diet supplied to intensively housed chickens must supply sufficient nutrients to enable the hen to produce an egg with a good shell thickness and good egg yolk colour. A high level of yellow maize, leaf or grass meal will ensure a good yolk colour. Calcium carbonate in some form (limestone or shell) must be supplied. This is either mixed in the ration or fed as a separate supplement on a free-choice basis. It is often quite practicable to have a separate container in a pen with shell or limestone inside. Fish meal with a high fish oil content fed in the diet can give fishy flavours to eggs produced by hens on those diets.

Incidence of Disease

The diseases Infectious Bronchitis (IB) and Newcastle Disease both affect egg quality. They cause the hens to lay eggs with misshapen shells and poor quality thick white. IB induces groove-like marks along the long axis of the eggshell.

Management Control Over the Laying Flock

In many developing countries, there is a belief that a rooster is necessary to stimulate hens to lay. This is not true. The presence of an active male causes the eggs to be laid as fertile eggs (containing an embryo chick), and this reduces the storage stability of the egg. Even after the male is removed, all eggs laid are fertile for up to six weeks because sperm is stored and released from specialised cavities in the hen's oviduct. If fertile eggs are in demand, then cocks should be placed with the hens. Non-fertilised eggs have a much longer shelf life than fertilised eggs and are more suitable for the market.

Dirty eggs can be reduced in number. For hens in deep litter systems, the nest box litter must be clean and replaced regularly. Frequent collection of eggs under any housing management system, and at least four times a day in the hot humid tropics, will reduce the incidence of dirty eggs.

Management Control Over Egg Handling

Temperature Control

The most effective way to preserve egg quality is to store eggs between 10 and 15°C during all handling, transport and marketing phases. Insulated containers and/or vehicles can maintain cool temperatures during long-distance transport.

Even an outer layer of straw in a basket will help. In hot weather, and where there is no cool storage system, eggs should be transported to market at least every third day. Eggs should never be left standing in the sun or in a very hot room. Air conditioning or even an electric fan is advised whenever practicable. However, as air conditioning has the negative effect of drying out the egg contents as well as the advantageous effect of cooling, wet sacks should be placed as curtains in the cool store to alleviate this dehydrating effect. If fans or air conditioning are not available, then shaded well-ventilated rooms or underground cellars should be used.

Treatment of Dirty Eggs

An egg's shell has a natural protective coating (cuticle) that resists the entrance of bacteria and retains moisture inside. Washing eggs with water removes this protection, and thus washed eggs should be eaten as soon as possible. Whether eggs are wet- or dry- cleaned, they should be sold separately from naturally clean eggs, as their storage life is shorter. The cuticle from the shell is a protein-fat substance, and the lack of a cuticle can therefore be detected with a simple ultraviolet (UV) lamp. Washed eggs (without a cuticle) are red in colour under UV-light, while a blue colour indicates that the cuticle is still present.

Dry Cleaning

Even with good flock management, some eggs will get dirty. The risks of allowing water to touch the shell have already been mentioned. Dry cleaning systems are preferred. Rubbing lightly with fine sandpaper or a rough cloth is better than wet cleaning. Cloth-backed sandpaper or emery paper can be wrapped around a block of foam rubber for dry cleaning by hand. Steel wool and nylon dishwashing or bathroom scrubbing aids are also quite suitable. Care should be taken not to remove too much of the protective cuticle layer which covers the shell. Only the dirty patches should be cleaned. There are also motor-driven dry-cleaners commercially available. The simplest model consists of a spinning wheel of foam rubber. A mixture of glue and sand is applied periodically to the foam wheel. The operator holds the egg against the spinning foam wheel to clean it.

Wet Cleaning

Washing of eggs is only suggested under very well-controlled conditions. The concern is to ensure that the washing water temperature (38 to 43°C) is never below that of the egg. This avoids the wash water being sucked into the egg through the shell pores by the action of the egg contents shrinking (as happens if the egg is in contact with cooler water). In addition, the washing machine must be able to monitor the detergent/sanitiser/disinfectant/antiseptic levels in the water to ensure that they are optimal. Only special types of non-tainting chemicals can be used. The water itself must be changed frequently. After washing, the shell should be pasteurised by dipping the eggs in water at 82°C for a few seconds, then dried quickly with warm air before packing. The eggs must also be clearly labelled as "washed". Washing done in this way is complex and expensive, and is therefore only justified in large operations, although even then it involves risks.

Egg Quality Grading

Interior quality

Candling

Opening the egg by breaking it is the only accurate way to fully check the interior quality. This can only be done on a limited sample basis. "Candling" can show some aspects of internal quality without breaking the shell. It consists of inspecting the egg in a beam of light strong enough to penetrate the shell and illuminate the contents. Various types of lamps can be used but the essential features are similar. An incandescent-type bulb of 25 to 50 watts is enclosed in a casing with light exiting through a round hole about 3 cm in diameter against which the egg is held and turned. The casing usually has another hole to provide light for the operator to see the egg container if the room is very dark. By rotating the hand-held egg close to the hole in the candler, the yolk and egg white quality can be estimated by their movement. Experienced operators can candle 24 eggs per minute. The main points to observe are summarised in the following paragraphs.

White

Egg white (albumin) characteristics showing good egg quality are thick albumin fullness and albumin transparency. When the thick albumin sack is strong and healthy, it is full and confines the yolk within the various layers of egg white. As the thick albumin sack deteriorates, its contents leak into the thin albumin cavity. The yolk then moves more freely, increasing the risk that it might touch the shell and be contaminated by microorganisms from outside the shell. A healthy albumin is also transparent. It can become discoloured or cloudy due to rot or overexposure to hot water (partial coagulation) in washing.

Yolk

Yolk characteristics showing good egg quality are confinement within the thick albumin, a small spherical shape, orange-yellow colour and the absence of spots. As described in the above paragraph, yolk confinement within the albumin protects the yolk from outside contaminants. A small spherical shape indicates a strong yolk membrane. When the egg is exposed to high temperatures and dehydration, the yolk deteriorates and grows larger and flatter. Consumers prefer yolks of orange-yellow colour without spots. Spots on the yolk can indicate: embryo development (reddish colour); blood from the hen's ovary and "meat" bits from the oviduct released during egg formation (red and brown, respectively); moulds (grey or black); or bacterial rots (blue, violet, green or red). Although consumers prefer yolks with no spots, the only spots that pose any health risk are mould and rot spots.

Air cell

Air cell characteristics showing good egg quality are small

size, shallow depth and fixed position at the blunt end of the egg. Small size and shallow depth indicate very little loss of moisture from the egg contents, which in turn indicates freshness (or that the eggs have been stored under good conditions). A fixed position at the blunt end of the egg indicates that the membranes surrounding the air cell have not been damaged.

There is usually a correlation between the depth of the air cell and other quality aspects. However an egg stored at high temperature and high humidity may show a good air cell depth (as the high humidity maintains the egg moisture) but it may have deteriorated otherwise (as a result of the high temperature).

Air cells can be deflated completely or become unfixed and mobile within the egg. The air cell can become filled with albumin if part of the inner shell membrane is broken. If the membrane is merely weakened, the air cell may move freely around the egg. These mobile air cells are often caused by transporting eggs on rough roads or by the egg being stored small end upwards. The egg could be otherwise quite fresh.

Shell quality

Before candling, eggshell quality is assessed, and eggs that are dirty, cracked, thin, rough or misshapen are processed accordingly (procedures regarding shell quality are addressed extensively in the above section on production factors affecting egg quality).

Egg Sale Options: Graded Size or Total Package Weight

Eggs can be sold by graded size or by total package weight. Selling by graded size involves weighing each egg individually and grading the eggs within certain weight ranges (commonly Small, Medium and Large). They are then packed in cartons of 10 or 12 eggs, and sold according to a price per graded size. Selling by total package weight involves packing the eggs without size grading, and selling the package according to a price per kilogram (like almost all other food products).

Consumers in the more developed countries are accustomed to buying eggs graded by size and boxed into cartons. Grading eggs by size requires complex machinery for grading and packaging, as well as monitoring and testing of all grading machines, and sample monitoring of the various grades at retail outlets.

In developing countries without the capital or administrative capacity to undertake such extensive monitoring tasks, the better option is to sell eggs by total package weight. If a market weighing scale is used to weigh foods such as rice or maize, then it can also be used to weigh eggs so that they can be sold by package weight. Selling eggs by package weight also simplifies the situation where standardisation of containers and grades has not yet been developed. It also makes price comparisons between different types of food items much easier for the consumer.

Eggs in most developing countries are sold by quantity rather than by weight, which penalises the producer of larger eggs. As local breeds of hens usually lay uniformly small eggs, this is not a significant problem. However, as the market grows and a demand develops for different sized eggs based on the availability of commercial hybrids (laying larger eggs) in peri-urban areas, the decision to sell eggs by graded size or by total package weight must be faced.

Egg Transport

The four concerns regarding egg transport are:

— Protection against mechanical damage, which can be achieved by avoiding excessive shaking, especially where roads are bad, and by using spring suspensions on bicycle carriers.

— Protection against poor egg handling, which can be achieved by providing convenient loading levels to make lifting easier.

— Protection against tainting odours.

— Protection against exposure to high temperatures in transport.

Egg Packing Methods

Eggs can be packed with a padding of rice husks, wheat chaff or chopped straw in firm-walled baskets or crates. This greatly reduces the risk of shell damage in transport. In Iran, long flat boxes, each containing about 1,000 eggs cushioned in chopped straw, are commonly used for the transport of eggs to the capital from a distance of up to 1 000 km. The boxes are transported in trucks over rough roads, but breakages seldom exceed five percent. The main difficulty with such systems is in standardising the number of eggs per container. Consignors and receivers will otherwise spend much time counting eggs and repacking to ensure that the correct number has been received for payment.

The standard type of transport egg packing container is the 30-egg tray, which is made of paper pulp holding six rows of five eggs each. The trays are stackable either when full or empty. A standard box of 360 eggs (30 dozen) is made up of two stacks, each comprised of six trays. Washable plastic reusable trays are also available. Cases are usually made of wood. Half-cases to hold 180 eggs (15 dozen) are also common and are usually made of corrugated cardboard.

Quality Preservation During Transport

Permissible temperature ranges depend on the duration of transport time. In Europe, the temperature recommendation for two to three day transport in refrigerated vehicles is between -1°C to 3°C. In developing countries, however, refrigerated vehicles are not widely available. Even when available, precautions are needed to avoid moisture condensation on eggs removed from the cool container to the warm moist air of the retailing environment.

Fans blowing air towards the eggs across a container of salt and ice is a cooling system that has been used in Pakistan for egg transport by rail for the 1,600 km trip from Peshawar to Karachi, where outside summer temperatures can range from 38°C to 47°C.

Refrigerated transport is expensive. In estimating the costs of establishing such a system, the volume of trade for refrigerated goods is an important consideration. The capital cost may be spread over five years to prepare the costing. Transport of other taint-compatible produce with the eggs should be considered; as should the prospect of back-loading with other goods, which may not necessarily require refrigeration.

Public transport such as rail or bus is the most common means of transport in developing countries. Awareness of the special needs for egg transport as addressed above will assist the operator in preserving the quality of the eggs no matter what the type of transport.

Egg Storage

All egg storage systems must meet the following requirements:

- — Water loss by evaporation to be minimised.
- — Mould and bacteria growth to be minimised.
- — Interior quality to be maintained (indicated by a good proportion of a thick white, a firm, rounded yolk and good flavour in both).

The first two requirements can be met (for storage periods of three to five months) by: coating eggs with oil or waterglass (sodium silicate); immersing eggs in limewater (calcium hydroxide solution); or putting eggs in dry storage (using such materials as bran, peat dust, soda lime, salt and wood ash). However, all three of the above requirements can only be met by refrigeration, which is the best storage method, if available.

Following below are descriptions of some of the traditional egg storage methods used in the absence of refrigeration. The first two systems rely on evaporative cooling, which is only effective in the hot dry tropics. The hot humid tropics do not allow sufficient evaporation to occur, and thus there is much less of a cooling effect. Where none of these storage systems can be used, there is no way to slow the inevitable drop in egg quality, and the eggs should therefore be transported to the consumer as quickly as possible.

Clay Pot

Eggs are placed in a clay pot buried in the ground up to its neck, in a shaded area. The pot is covered tightly so that no water gets into the pot. The ground around the pot is watered, but without leaving puddles of water. Straw or a mat is placed in the pot to cushion the eggs and to keep them above any water that seeps into the container. The eggs are put in the pot as soon as they are collected, and covered with a cloth and damp straw. Due to the evaporative cooling effect, the inside of the pot is often five to six Celsius degrees cooler than the outside air temperature.

A variation of this method, used in the Sudan, is to bury an earthenware pot in the ground to half its height. A 7 cm layer of mixed sand and clay is packed around the pot up to its neck, and kept wet by sprinkling water on it. The inside of the pot is lined with grass. The eggs inside are covered with a thin cloth to allow air circulation. Evaporative cooling in Sudan's hot dry climate often reduces the egg temperature to up to eight-Celsius degrees below that of the air outside. Eggs are turned daily to prevent the yolks touching the shell, which would accelerate the decaying process.

Wet Sack Cooler

This is another method utilising the evaporative cooling principle. The sack material is kept wet by having a tray of water above the hanging sack, into which the neck-edge of the sack material is dipped, keeping the sack wet. A slightly more sophisticated system uses perforated pipes connected to a water tank. To prevent mould formation, the sacking is pre-soaked in a solution of copper sulphate ($CuSO_4$), using 60 g of crystals in four litres of water.

Oil Coating

A thin film of oil on an eggshell fills its pores and reduces evaporation and thus spoilage of the egg contents. Using a wire basket, the eggs are dipped into slightly heated oil, about 11°C warmer than the eggs. Special odourless, colourless, low viscosity mineral oils can be used. If these are not available, then any light mineral oil or almost any cooking oil that doesn't easily turn rancid serves the purpose. To reuse the oil, it is cleaned through a filter and heated to 116°C to sterilise it. Four litres of oil coats about 7,000 eggs. Oiled eggs last for at least three weeks (longer if kept at 10 °C, or less at temperatures above 21°C). For high temperature storage, eggs should be oiled four to six hours after laying.

Waterglass Paste

Waterglass is a paste or ointment of sodium silicate in water. It is rubbed onto the hands and then the egg is rolled between the two waterglass-coated hands to transfer a waterproof coating of waterglass paste to the eggshell.

Waterglass Solution

For 100 eggs, a 25-litre pot or jar is used, and 5.3 litres of previously boiled (and then cooled) water are mixed with 0.5 litres of waterglass. The eggs are placed in the pot and covered with the waterglass solution. The pot is covered and kept in a cool, shaded place. The eggs keep for one to six months.

Limewater Solution

Limewater is a solution of calcium hydroxide [$Ca(OH)_2$], a mild alkali. The main ingredient is burnt lime (also known as quicklime). The chemical name of this is calcium oxide (CaO). It is also known as choon in Bangladesh, and is a common ingredient of the betel nut mixture chewed by people in many tropical countries. Calcium oxide is made by burning limestone ($CaCO_3$) in a hot fire. Carbon dioxide (CO_2) is driven off from the limestone, leaving CaO behind as a white powder. Dissolving this calcium oxide in water makes limewater. The resultant solution of calcium hydroxide is only partly soluble, and the insoluble portion will settle to the bottom of the container.

Six litres of limewater is made by stirring 2.3 kg of calcium oxide into six litres of boiled (then cooled) water. It is allowed to stand overnight so that the insoluble portion settles. The eggs and the clear part of the limewater solution are placed in a pot, covered and kept cool. The eggs last more than a month. In the years prior to 1970, eggs were commonly transported from Bangladesh (formerly East Pakistan) to Pakistan (formerly West Pakistan) on a train journey of about a month, in high temperatures. The eggs were stored in earthenware jars containing limewater and maintained their quality well.

Hot Water Immersion

Immersion in hot water for carefully controlled lengths of time has a pasteurising effect, which kills the embryo in fertile eggs, destroys some of the bacteria on the shell and stabilises the quality of the egg white. The difficulty is to achieve this without coagulating some of the egg white. Equivalent effects are achieved with any of the following temperature-time combinations. This method requires special equipment and supervision.

Salt and Wet Clay or Ashes

Eggs are coated in a mixture of salt and wet clay or ashes which allows them to keep for one month. This method has been practised for centuries in China.

Cooked Rice and Salt

Eggs are covered with a mixture of cooked rice and salt, which allows them to keep for six months. This method has been practised for centuries in China.

Lime, Salt, Wood Ashes and Tea

Eggs are covered with a layer of lime, salt and wood ashes mixed with a tea infusion, which allows them to keep for several years. This method has been practised for centuries in China.

Today, there are three leading systems for producing poultry outdoors on pasture. There are significant differences between the systems but sometimes the systems are confused with each other. Here is a short course in range (short-grass pasture) poultry nomenclature.

Free-range Poultry Production and Marketing

Free-Range is a non-confinement system that uses a perimeter fence to deter predators. As defined, this means at least 150' between the skid house and the closest fence. A variation of this system, known as DayRange, uses an Electronet portable fence to keep the birds safe from dogs and coyotes during daylight hours. Up to 300 birds are housed on an 8′x 18' wooden skid shelter that is towed to new pasture once every week or so as needed. Birds return to this shelter on their own each evening and are confined for predator protection.

During the day the birds are let out and the skid provides shelter from sun and rain when needed. The large-scale access to pasture combined with the low stocking rate (400 chickens or 100 turkeys per acre) allows the birds plenty of area to exercise and deposit manure. As a result, free-range birds develop excellent muscle tone. Since the muscle is what we eat, this development is very important. Combined with proper aging after slaughter meat quality is firm but smooth—second to none.

The free-range system uses bulk feeders and float-valve waterers connected to plastic pasture waterlines. This cuts chore time dramatically as water is never hand carried to the flock and feeders are usually sized to need refilling once per week. Daily chores involve opening the skid houses in the morning, checking to see that the systems are operating properly and the birds are healthy, and closing them up a dusk.

One of the first questions to come from prospective producers is about predation. Because most predation occurs at night, when the birds are totally enclosed on the skid, which has a sturdy wood floor and 1" poultry mesh sides, there is normally no problem. Hawks occasionally steal an inattentive bird, but this is rare. Most birds are extremely wary of sky activity.

Dogs and coyotes are another matter. A good perimeter fence solves most of the problems, but if there are severe dog or coyote problems in your area you may wish to investigate electric polywire fence (four wires on step-in posts keep most predators at bay) or Premier's electric Poultry Netting.

A real plus for the diversified farmer using the free-range system is the ability to "save" manure nutrients and make compost right on the skid itself. Producers using the Salatin system, have good fertility in the pasture, but are not able to save and transport the nutrients to other fields. The free-range system is a large "farm-scale" system suitable for practitioners raising 500 - 20,000 chickens per year.

Pastured Poultry, as researched and taught by Virginia farmer, Joel Salatin, is a confinement system with a grass floor. Using portable pens approximately 8 x 10 feet in size, this popular system is a big improvement over the broiler houses used by companies such as Tyson and Perdue, but it is a confinement system just the same. The pens, each containing about 80 chickens, are moved by hand but the lightweight construction means they can occasionally be blown over by the wind.

In hot climates, birds can suffer heat stroke on calm days. The birds have a limited space for exercise and manure that space heavily. Therefore, the pens must be moved twice daily, a chore not always pleasant, especially after a heavy rain. The birds benefit from sunlight, bugs and grubs, and get minerals from the soil, but muscle tone is very different from birds allowed to free-range. The Pastured Poultry System does protect against hawks, but is actually less protective at night against skunks, foxes, opossums and raccoons since it has no floor. The Salatin system is very labor intensive involving daily movement of pens and delivery of feed and water, but is well suited for those with limited space, those desiring to raise less than 1000 birds per year or persons who must work away from the farm during the day.

A third system, the Chicken Tractor was developed by Andy Lee and is a useful system for raising 50 or so birds for home use. By placing these pens in the garden, soil is tilled and manure can be placed exactly where desired. This is not a commercial sized system, and is also a confinement system. A recent refinement of the Chicken Tractor is the addition of a pop-hole door to allow the birds to range at least part of the day.

Producers using both the "pastured poultry" and "chicken tractor" systems have reported leg problems with chickens. While part of this problem can be eliminated by using the heavier strains of Cornish Cross chickens designed to be raised to roaster size, most leg problems are caused by the pens being pulled over the legs of the chickens when the birds are being moved to new pasture. Neither of these two systems are suitable for turkey production.

Some farmers using the latter two systems advertise their chickens as "free-range", but under European rules they would not qualify as the birds are too tightly confined. Whichever of these systems you choose, to make a living wage you will need to ask at least $2.00 per pound dressed weight. Right now consumers are way ahead of farmers in demanding "grass-fed" poultry. If you are looking for a consistent income producing livestock enterprise for your farm, one that easily integrates with cattle, sheep or vegetable production, consider adding poultry to your operation.

Profitable Poultry

Raising poultry on pasture isn't exactly new. Most broilers, layers and other domesticated fowl were raised outdoors before the advent of the now-dominant confinement method in the late 1950s. Since then, large corporations have become the primary producers of poultry in the United States developing "vertically integrated" practices that allow them to capture nearly 100 percent of the multi-billion dollar annual market. Today, vertically integrated corporations control almost every aspect of how broilers and eggs are produced, processed and sold.

Individual farmers still participate, but as contractors who agree to meet standards that usually include furnishing climate-controlled confinement houses to hold 25,000 birds or more. Each house costs as much as $140,000. Poultry companies usually supply farmers with chicks and feed needed to bring them to market weight in about seven weeks. They also supply subtherapeutic antibiotics to prevent disease, growth promotants for faster weight gain and drugs to control coccidiosis, common in concentrated operations.

The vertically integrated corporations then typically manage the slaughtering and packaging process, paying contract farmers by the bird, with feed and heating costs factored into the equation. The system has helped make chicken a low-cost staple for American consumers.

But some farmers and consumers question whether, in the process of achieving that efficiency, values they consider important – autonomy and independence for farmers, the welfare of the flocks, and the taste and quality of their meat and eggs – have been lost. To meet a growing niche for poultry raised differently, a number of growers are choosing to raise birds in alternative ways, most of them reliant upon pasture.

Poultry System Options

In the early 1990s, Virginia farmer Joel Salatin published a book detailing a new system to compete for the small but growing niche of consumers who want to buy poultry raised outside the corporate system. His Pastured Poultry Profits, 10,000 of which have sold, explains the innovations Salatin made to the old practice of allowing poultry to range free around the barn lot. It lays out production strategies alongside promises that readers who follow his methods can net $25,000 in only six months on 20 acres.

Chickens are raised in floorless, 10′ x 12′ x 2′ pens containing 75 to 90 broilers. Producers move the pens daily to fresh pasture. While receiving exercise and fresh air foraging for plants and insects, the chickens drop manure that adds fertility to the soil. Producers buy day-old chicks between April and October, then move them from brooders onto pasture after a few weeks.

According to many, Salatin's book sparked a renewed interest in raising poultry on pasture. The book details how to brood chicks, rear birds in pens, slaughter, dress and package the birds, process eggs, and sell poultry products. Salatin has since begun to work directly with others to pass along his experiences and ideas, holding field days and speaking frequently at conferences.

Marketing Options

The experience of practically every range poultry producer bears this out: Marketing your product will take as much time and energy as the actual task of raising and processing your product.

Marketing Tips

Newspaper Stories. Marketing is all about capitalising on advantages. The key lies in what some call "relationship marketing." Farmers who have received ink in newspapers or magazines report that when people read about their product – and the philosophy and practices behind pasture-based poultry systems

Pre-Orders. David Bosle of central Nebraska prints a newsletter every winter for his customer list of close to 300. He includes a self-addressed, stamped envelope to take orders, by month, for the season. "That's a must," he said. "The biggest cost is to get the customer on the belt, and once you've got them there, it's stupid to let them fall off." Bosle takes advantage of the short growing time for chickens and clusters his flocks around spring, summer and fall holidays, including Memorial Day and Labor Day. With the pre-ordering system, he generally sells his birds prior to growing them.

Samples. Robin Way not only praises the virtues of investing in a colorful, easy-to-spot farm sign, but also recommends giving out free meat. "If they take the trouble to drive down our lane, I'll give people freebies," she said.

"Maybe they'll never show up again, but may be they'll be one of our best customers."

When they first starting raising poultry on pasture, the Ways would bring chickens to auctions and meetings and make donations. All of the meat was accompanied by their farm business card.

Selling with Other Products. Delahanty, the New Mexico grower, markets his organic meat under a "Real Chicken" brand that commands varying premium prices – as high as $5 per pound at some upscale groceries in Albuquerque and Santa Fe. Next, he plans to sell organic vegetables he expects will flourish in the manure-rich soil aided by his flocks.

The Extra Healthy Egg

Some producers are trying to capitalise on the ability to enrich eggs with omega-3 fatty acids, which lower cholesterol and thus have been linked to reduced risk of heart disease in humans. Flax, commonly grown as an oilseed, can be added to hen rations at about 15 percent.

Niches Within a Niche

Thirty-one percent of the respondents to a recent survey raise turkeys along with pastured broilers. Sixty-nine percent raise layers. Many also report raising varieties of poultry other than chickens and turkeys, including ducks, guinea fowl and pheasant hens. Some market two-year-old stewing hens.

Specialty fowl such as ducks can be raised with as little effort as is required for broilers and layers, but can bring in much more money per pound. Their rareness also tends to make the job of marketing easier. Restaurants offer good markets for exotic fowl, and if regulations in your state allow direct sales to restaurants, it's worth contacting the chefs at every upscale establishment in your area.

Label Rouge: Pasture-based Poultry Production in France

Pasture-raised poultry is the leading product in a programme in France called *Label Rouge*. This programme provides premium products to consumers, increases farmer income, and strengthens rural development. It consists of many regional producer-oriented alliances, called filieres, which produce and market their own branded products under a common label. A third-party certification programme ensures that strict standards are being followed. Other countries are beginning to take note of the Label Rouge programme.

In the U.S., a grassroots "pastured poultry" movement has been growing since the early 1990s. Poultry raised on pasture are processed on-farm and direct marketed, creating supplemental income on small diversified family farms.

Label Rouge is divided into 2 sections:

— Production

— Organisation

Production

Label Rouge began 40 years ago as a grassroots movement led by visionary farmers. As poultry became more industrialised after World War II, demand grew in France for the taste of traditionally raised farm chickens. Label Rouge performance has been "stunning" and now accounts for 30% of poultry sales to the public, in spite of its high price—twice the price of conventional poultry.

The *Label Rouge* programme focuses on high-quality products, mainly meat, with poultry as the flagship product. It emphasises quality attributes such as taste and food safety, and free-range production practices. The average consumer can note a positive difference in taste between *Label Rouge* poultry and conventional poultry—in fact, regular taste-testing is a certification requirement to prove that these products are "vividly distinguishable" from conventional poultry.

The main reason for the superior taste is considered to be the use of slow-growing birds instead of the fast-growing birds used in the conventional industry. The slow-growing birds are from specialty "rustic" genetic stock and are harvested close to sexual maturity. The meat is flavorful and firm, but not tough.

Standards

Strict and comprehensive standards ensure quality. Independent third-party certifying organisations ensure that standards are being followed. Inspection occurs once per flock, twice per year for feedmills, monthly for processing plants, and twice per year for hatcheries. Each visit includes bacteriology tests and process control inspections. There are five taste tests per year.

The standards are a baseline that many *Label Rouge filieres* surpass. For example, some groups:

- — Use dividers in the house to divide flocks into smaller flocks.
- — Require tree and bush plantings to integrate the house into the countryside as well as provide shade on pasture.
- — Use smaller, portable houses.
- — Do not permit pesticide use on the range.
- — Require grit and whole grains to improve gut health.
- — Maintain a constant ration to keep the taste of the birds constant, not changing it when other ingredients are less expensive.

Besides broilers, standards also exist for layers, turkeys, ducks, geese, guineafowl, and capons. Layers require double yards (rested in rotation) since they are on the range longer than broilers. The standards are available (in French). There are also Label Rouge ham, sausage, eggs, rabbit, and cheese products. As is evident from this discussion of standards, a certification programme can permit much broader production claims than a mere definition can. In addition, production claims can be verified by the consumer.

Case Studies: Lands filiere and Lou filiere

The journey from a grassroots movement to an industry can

be seen by studying two different *filieres*. The *Label Rouge* movement began in the '60s in the Southwest of France, in the forested Landes region. Landes poultry are still known for being raised in a pine forest, using small portable housing called "Marensines."

The size of the buildings ranges from 16' x 16' (270 sq. ft.) to 20' x 33' (645 sq. ft.). Older houses were built of wood; new ones are metal. In a dense forest, the smaller houses are used to fit between the trees. Litter is spread in the houses, which are floorless. Brooding is done in the houses with portable gas brooders. Part of the feed is kept outside to help train birds to go out.

The houses are moved after every grow-out (three times per year) and have knobs where wheels can be attached and towed by tractor. The houses are sometimes placed beside cornfields so that birds can benefit from shade and forage for insects. Often, only three sites on the farm are used in rotation. The sites have a water line or producers fill barrels, which feed water troughs, every couple of days.

George Berbille invented the portable "Marensine" system 40 years ago and is considered the father of range poultry production in France. His farm is in the Southwest in the Landes *filiere*. He is now elderly and has lost a leg to a combine but still raises corn and poultry (50,000 birds per year by himself). He has 20 small houses, which take two days to move with the help of three people. (They are dismantled before moving.) He also has an on-farm feedmill and mixes feed daily for his use and for sale.

Although *Label Rouge* production began in the Landes region, it was the Loue *filiere* that was instrumental in making it a viable industry. *Label Rouge* did not grow as an industry until the product became widely available at supermarkets.

Loue is now the largest *filiere* and represents the typical production system used—a small fixed house and yard. The house has automated feeding and watering equipment, and chicks are brooded in it. The house has several popholes that allow access to the range. There are shade bushes planted in the yard as well as tree plantings. The yard immediately outside the house is dirt. Since specialty "rustic" genetics are used, the birds forage well beyond the house and have access to grassy pasture. Feed and water are also provided outside. The Loue Web site is available in French only but has many informative pictures.

It is not necessary to use stationary housing in order to build a national industry. U.S. pastured-poultry producers are keenly interested in pasture rotation and use portable houses; in adapting *Label Rouge* features, U.S. producers are more likely to promote portable housing than stationary housing. However, small specialty companies may adapt a stationary system. Production systems are an excellent way to differentiate between companies in the marketplace, as long as the basic standards are followed to market under a common label.

Label Rouge birds are usually produced on diversified farms where they are integrated with other livestock and grain production. Bird rations are supplemented with whole grains from the farm; litter from the house is spread on the fields. Poultry may bring in 50% of the farm income.

Although no fence is used in either the Loue or Landes *filiere*, the loss to predators is only 1% (the U.S., however, has more predator pressure). European Union (EU) definitions differentiate fenced and unfenced production systems:

— Fenced: "raised in open air"

— Unfenced: "raised in total freedom"

The *Label Rouge* programme permits both. The European Union specification 1538-91 defines *Label Rouge* as "traditional free-range poultry."

Genetics

Slow-growing birds are key in *Label Rouge* production—birds grow to 5 lbs. in 12 weeks. In comparison, the fast-growing broilers (Cornish cross) of the conventional industry reach 5 lbs. in 6-7 weeks. Not only does slow growth allow the organs, muscle, and bones to grow in harmony, it also results in a more flavorful meat. The carcass is generally more elongated and has a smaller breast and larger legs than conventional carcasses. In addition, slower-growing breeds are more suited to outdoor production than Cornish cross.

In Europe the slow-growing genetics are mainly supplied by the poultry breeding companies SASSO, and Hubbard-ISA. They do not sell the actual broiler chicks, but rather the parents; however, many pastured poultry producers have hatching capability. SASSO's typical *Label Rouge* cross is T44N male x SA51 female (using a different male—the T44NI—results in white underfeathers in the offspring). A typical Hubbard-ISA cross is S77N male x JA57 female. Broilers from both of these crosses will have red feathers, yellow shanks, thin skin, and a naked neck. Other parents are available for broilers with white feathers and skin, black feathers, barred, non-naked neck, etc. or for faster growth.

SASSO and Hubbard-ISA genetics of this type are not available in the U.S. However, a U.S. company called Rainbow Breeding Company is developing similar genetics and offers Free Range (FR) Broiler parents. FR Broiler chicks (day-olds) are also available. Male chicks are regularly available; female chicks are available only sometimes since they are used more in breeding; females grow at 85% the rate of the males.

Medium-growing genetics are also of interest for U.S. pastured-poultry producers. Producer Tim Shell is developing a broiler called CornDel which is a Delaware/Cornish cross combination. Grow-out is 9 weeks. According to Shell, the chicks can be used for parents if feed is restricted so they will not get too fat. Availability is limited. Henry Noll offers a Silver Cross that grows to 5 lbs. in 9 weeks. Joe Cebe, Sr., offers a Cebe Red and Cebe Black meat variety that grows to 5 lbs. in 9-10 weeks.

Redbro is a Hubbard-ISA Shaver product that happens to be available currently in the U.S. via a Canadian company that imports parents from France. It grows out in 9-10 weeks. Jerry Srednicki at a Connecticut hatchery ships day-old chicks. There is interest in using standard heirloom chicken breeds for gourmet poultry production, but in general, heirloom breeds have not yet been selected for commercial production and the carcass will be very small at 12 weeks.

Turkeys are native to the Americas and there are several slow-growing breeds available. These are naturally-mating turkeys and do not require artificial insemination. Some have geographical ties to the regions in which they were developed (e.g., the Bourbon Red is from Kentucky and the Narragansett is from Massachusetts). Mike Walters Hatchery, offers eight heirloom turkey breeds, some of which have been selected for commercial production.

Health

The use of slow-growing genetics and the low-density *Label Rouge* production system offer distinct health advantages—ascites, leg problems, and sudden death are minimal, and birds have good immunity. Mortality for conventional broilers in France is 6% during a 6-week grow-out; it is half that for *Label Rouge* production (3%) even during a much longer grow-out (12 weeks).

Since *Label Rouge* birds have a longer life, they have a different vaccination schedule than conventional broilers. For example, in France, conventional broilers are not vaccinated for Marek's Disease; *Label Rouge* broilers *are* vaccinated. *Label Rouge* birds are generally vaccinated for coccidiosis and given dewormers in the feed. Probiotics are used; antibiotics can be used only if prescribed by a veterinarian. Regular biosecurity on the farm is important—footbaths are used at the entryways to houses and visitors must wear protective clothing.

Since France has a mild climate, birds are raised outdoors year-round. However, the mean number of flocks per year is only 3.2 since the grow-out is long and there is a long downtime required between flocks for proper sanitation and pasture rest.

Feeding

A low-protein and low-calorie diet is used for slow-growing birds. Whereas typical fast-growing Cornish cross rations in the industry start at 22% crude protein and finish at 17% protein, *Label Rouge* rations start at only 20% protein and finish at 15%. According to Jeff Mattocks of Fertrell, pastured poultry producers in the U.S. often use only one ration of 19% protein. A low-protein ration is used to slow down the rapid growth of Cornish Cross. This type of ration could easily be used for slow-growing genetics.

All meat meal is banned from livestock feed in Europe. Even fishmeal is not permitted in *Label Rouge* production because it could be confused with meat meal. Only vegetable fat is permitted and no genetically engineered crops can be used in feed.

Some feedmills in France are dedicated to the eradication of Salmonella in feed. At Landal, a feedmill in the Southwest that supplies feed to the Landes company, entering trucks must be disinfected, and high heat is used during milling to kill pathogens.

Processing

Some processing plants exclusively process *Label Rouge* products; for others, *Label Rouge* is only a percentage of their work. There are several large automated *Label Rouge* plants (for example, Fermier Landes processes 200,000 birds per week), as well as small ones. There are many quality-control points during *Label Rouge* processing to ensure a high-quality carcass. Processing plants in France cool carcasses by air-chilling instead of immersion-chilling. (In immersion-chilling, the carcasses soak up water.) A soft scald is used instead of the hard scald typical in the U.S. (A soft scald uses a lower temperature for a longer time than a hard scald and keeps the skin intact.)

Although ready-to-cook products are the most common, a variety of dressing methods are used. In the *eflilee* style, the bird is eviscerated but the crop, head, and feet are left intact. Corn finishing is a part of this presentation—the customer should be able to feel whole grains still in the crop. Birds dressed in this style are slaughtered and eviscerated manually.

Processing plants may also handle a variety of species. For example, *Fermier Landes* processes chickens, guinea fowl, cockerels, and rabbits, as well as capons and turkeys for Christmas. Although it is a large plant, they can put together small custom orders for butchers and other clients.

Organisation

The key to the *Label Rouge* system is the supply chain. Grassroots pastured-poultry producers in the U.S. are largely independent and may be particularly interested in supply chain structure and the benefits that coordination can offer. Coordination of the supply chain offers a number of benefits, including coordination of the stages of production, lower costs, ability to reduce pathogens throughout, and complete traceability.

Supply Chain Structure

Filieres

The *filiere* is centered around a group of producers and associates called the "quality group," which holds managerial responsibility for the *filiere*. This body—not the processor, input supplier, or distributor—retains control of the brand and makes the pricing, marketing, and advertising decisions. This allows for a balance of power among producers and other players in the supply chain. Farmers earn more per bird from *Label Rouge* production than they do from being contract growers in the conventional poultry industry; however, conventional contract growers raise more birds per year. Since farmers are represented in the quality group, they have input into the returns they get on their birds and the number of flocks they raise.

Deciding on the margin that goes to the farmer is a crucial point for each supply chain. The *filiere's* quality group reviews the market prices—for both the production inputs and the final products—frequently, and adjusts the farmers' margin and the price to the consumer as needed. Risks and rewards are shared by all players in the chain.

Each sector incurs costs from being part of the supply chain, including costs associated with certification, such as time required to keep records. Each sector may also have to contribute to a check-off that funds the group's marketing and other activities. Thus each sector must have some incentive to participate, and the rewards must outweigh the costs.

There is no one model for *filiere* structure, and there is a lot of crossing-over within and among the *filieres* in the *Label Rouge* system. For example, a processing plant may serve more than one producer group; a breeding company may sell breeding stock to all the *filieres*.

Benefits of Coordination

A coordinated supply chain helps lower production costs by improving cost efficiencies (bulk purchases, etc.) and coordinating the stages of production. It also allows complete traceability of the final product. The *Label Rouge* traceback system is so detailed that batches of birds can be traced from the consumer to their grandparents' flocks. A code on the package indicates the farm, the origin of the chicks, the processing plant, and so on. Wing-banded birds can even be traced individually.

Coordination permits quality control throughout the supply chain. If there are broken wings in the picker, where did they come from—the picker or the farm? It helps cooperating growers produce a consistent product. A coordinated supply chain also provides an excellent opportunity for addressing food safety. In *Label Rouge* production, a pathogen-reduction or HACCP-type programme is applied not only at the processing level (as required in the U.S.) but throughout the entire supply chain. For example, breeding flocks, hatcheries, feed mills, farms, processing plants, and transportation are all monitored for Salmonella and other pathogens. *Label Rouge* has an excellent food safety record—only 3% of *Label Rouge* carcasses are contaminated with Salmonella. Sales of *Label Rouge* products have risen in the wake of several BSE or "mad cow" food scares, since consumers are assured that no animal by-products have been fed in the programme.

Another benefit of coordination is that *Label Rouge* provides technical support and trouble-shooting assistance to producers, which is particularly helpful for health issues such as a vaccination programme and disease diagnosis.

There is not a lot of public production research on *Label Rouge* production. Companies do research in their own farms or the farms of their producers, but there is not a mechanism to share this information.

Marketing

The *Label Rouge filieres* sell branded products that are strongly tied to regional areas and have their own images. For example, the Landes *filiere* markets the image of chickens ranging free in the pine forests along the Atlantic coast. There may be several regional brands competing in a supermarket.

In France, there is a concept of "terroir"—that taste comes from the earth. Certain regions are associated with certain tastes. In fact, most of the *Label Rouge* brands are names of French cities or regions. Such an "appellation" system is already well known with wines (e.g., champagne is from the Champagne region). Geographic restriction is important to prevent large-scale copying, and brand ownership keeps the brand names in the farmers' hands.

Most *Label Rouge* products are sold whole but the amount of cut-up is increasing. About 135 million birds are produced each year, and 15 million are cut-up. *Label Rouge* also offers an organic product.

Organisational Support

The *Label Rouge* system has strong organisational support both in the government and in the private sector. The French government is committed to strengthening rural development. Government agencies are responsible for setting and maintaining certification standards (CNLC), accrediting the certifiers (COFRAC), and protecting against label infringement (CERQUA). CERQUA protects *Label Rouge* from being copied by store brands or private labels.

To obtain a label, a *filiere* submits a *cahier de charge* to the CNLC. This document is a very detailed business plan, or code of practice, that details the production, processing, and monitoring process. It can take several months to get approval.

SYNALAF is an industry organisation that collects a check-off from the sale of each bird to conduct national consumer-education campaigns about the benefits of *Label Rouge* poultry. SYNALAF represents 38 filieres, which include 6,000 farmers—about half the poultry growers in France. Public education is key to the high premiums paid for *Label Rouge* products. Strong consumer organisations are involved in the development of standards and certification. This results in standards that are responsive to consumer interests, such as a recent ban on the use of genetically modified organisms (GMOs).

The *Label Rouge* system is complex but has built good working relationships with producers, consumers, and government that position family farmers to be economically sustainable. *Label Rouge* is farmer-created, consumer-driven, and government-supported.

Interest in labeling is growing in the U.S., where Certified Organic is currently one of the best-known labels. "Quality labels" like *Label Rouge* provide information to consumers on product attributes such as taste, health benefits, and nutrition, as well as on social issues such as support of local farms. They can also provide information on ecologically sound production practices and other factors related to sustainability.

Label Rouge Work in the U.S.

The University of Illinois has a project to promote the production and marketing of gourmet chickens in Illinois. In 1999, graduate student Francois Paybou, working under the direction of agricultural economics professor Randall

Westgren, carried out technical and economic feasibility studies for adopting the *Label Rouge* system in Illinois. Paybou determined feasibility to be positive, but considered the lack of an economical supply of French genetics and air-chill processing plants in the U.S. to be problematic. Another student, Amy Heady, did a market feasibility study to analyse consumer demand. She found that adoption of a *Label Rouge* system was still premature and too expensive given the current retail market. Feasibility may be greater in the restaurant market, where Chicago chefs, at least, are willing to pay $1.50 to $2.00 per lb.

Entrepreneur David Wilson started a *Label Rouge*-type business venture in the early '90s. He became interested in *Label Rouge* by talking to chefs who wanted a premium bird. He imported specialty genetics from France, calling them "La Belle Rouge," and contracted with growers in Kentucky and North Carolina who had older chicken houses on their farms. He followed *Label Rouge* requirements for flock size and density; the total range space was usually a couple of acres around the house. The broiler operation was year-round, but the birds did not go outdoors when the temperature dropped below 40°F, making a 90-day window in Kentucky in which the birds stayed indoors. During this time, stocking density was reduced and alfalfa was added to the feed. Grow-out was 12 weeks. Birds were shipped to an air-chill processing plant in Kentucky (no longer in operation). The meat was sold in Kentucky and North Carolina. According to Paybou, the Wilson business failed because investors rather than a farmer group were in control. They did not fully understand the system—the failure was not due to problems with the product or market.

More recently, the National Center for Appropriate Technology (NCAT) was funded by the USDA Foreign Agriculture Service's Scientific Cooperative Research Programme to travel to France to gather technical information about poultry produced under *Label Rouge*, to disseminate the information, to develop French contacts, and facilitate drafting of production standards.

Other French Labelling Programmes

Label Rouge is only one of four major labeling programmes in France. The other programmes include Appelation D'Origine Controlee (AOC), Organic, and Certificate of Conformity. All four complement each other well.

The major labeling programmes in France complement each other, reducing consumer confusion.

The *Appellation D'Orgine Controlee* (AOC) label is probably the best-known. As mentioned earlier, this programme reserves brand names for a certain region—for example, "champagne" refers to sparkling wines produced in the Champagne region of France. The product may not be reproduced outside of its region. Such labels are most often used for wine and cheese products, but there is a poultry product called "poulet de Bresse" that can be raised only in the Bresse region. These birds are known for being finished on milk.

The European Union actually recognises two types of geographic protection of agricultural products: Protected Designation of Origin (PDO) and Protected Geographic Indication (PGI). The PDO refers to the AOC label. PGI is less strict and refers to the *Label Rouge* system. A Web site from the European Union discusses quality labels.

The French Organic standards for poultry production are based on the European Union requirements but are stricter. Grow-out is a long 99 days. The Organic market in France is not as well-developed as in the U.S. because of competition from *Label Rouge*. Organic poultry products cost four times as much as conventional products, whereas *Label Rouge* products cost only twice as much. ATTRA can provide more information on organic poultry production.

The Certificate of Conformity programme is a quality-control label that is relatively industry-friendly. Medium-growth genetics are used. A fast-growing male is crossed with a slow-growing *Label Rouge* female to obtain a 5-pound bird in 56 days. Natural feeding is required, but access to the outdoors is not. This programme certifies the process used but does not have taste tests.

Opportunities

Label Rouge-type poultry production is an opportunity well-suited to the grassroots pastured-poultry movement in the U.S., as well as small specialty poultry companies. However, raising slow-growing broilers to 12 weeks costs more than raising fast-growing broilers to 8 weeks. Many grassroots pastured-poultry producers market directly to consumers on the farm or at farmers' markets, and their customers may not be willing to pay the higher price.

Small poultry companies or networks that serve larger markets or specialty markets may more readily find customers who are willing to pay extra for a pasture-raised, gourmet-type bird. Coordinated networks could keep the products at an affordable price by means of fine-tuned production systems and cost efficiencies. In France, consumers from all economic levels buy specialty poultry; not just wealthy consumers. The future development of a certification programme will be important for consumer education, which will in turn help build demand.

Large companies may find a certification programme similar to the "Certificate of Conformity" to be a more attractive opportunity than *Label Rouge*. In *Label Rouge*-type production, the flock size is limited to about 16,000 birds on one farm (in four small houses), and there are fewer flocks per year because of the long grow-out period. *Label Rouge* is designed for regional rather than national markets.

The outdoor production systems used in *Label Rouge* are more adapted to small diversified family farms than to large companies. Allan Nation, editor of *The Stockman Grass Farmer* and a U.S. visionary in the field of sustainable agriculture, found during his European travels that it can be a marketing advantage to family farmers to produce something that is "hard to produce."

References

Beck-Chenoweth, H., "Free-range poultry production & marketing: A guide to raising, processing, and marketing premium quality chicken, turkey & eggs", Back Forty Books, Creola, Ohio, 1997.

Beck-Chenoweth, H., "Free-range, pastured poultry, chicken tractor—What's the difference?" In: *Free-Range Poultry. Web Site.* Free-Range Poultry Production and Marketing, Creola, Ohio, 2001.

Buckels, C.G., "Midwest poultry producers share success stories", *Small Farm Today*; May 2003.

Cicero, K., "Homes on the range: Portable poultry pens are proliferating", *The New Farm*, May/June 1995.

Bibliography

Aboud, A.A.O., Owen, E., Reed, J.D. and McAllan, A.B., "Feeding sorghum stover to Ethiopian sheep: effect of stover variety and amount offered on growth, intake and selection", *Animal Production,* 1990.

Acharya, R.M. & A. Kumar, "Performance of local birds in South Asia", *Indian poultry industry yearbook,* 7th Edition, 1984.

Adene, D. F., "Pre-epidemic and epizootiologic features of avian influenza in Nigeria", *Western Poultry Disease Conference: Proceedings of the Western Poultry Disease Conference,* 2007.

Ahmed, N., *Backyard Poultry feeding systems in Bangladesh,* Asian Livestock XIX, 1994.

Aini, I., "Indigenous chicken production in south-east Asia", *World Poultry Science Journal,* 1990.

Aklobess, K.K., "Smallholder rural poultry production in Togo", In *Proceedings. CTA Seminaron Smallholder Rural Poultry Production, Thessaloniki. Greece 9–13 October* 1990.

Alders, R.G., "Women and village poultry: an old tradition returns to the fore", *African Network for Rural Poultry. Newsletter,* 1997.

Alvarez, F.J. and Preston, T.R., "Studies on urea utilization in sugar cane diets: effect of level", *Tropical Animal Production,* 1976.

Andreasen, C. B., "Training resources for prevention and intervention strategies during avian influenza", *Western Poultry Disease Conference: Proceedings of the Western Poultry Disease Conference,* 2007.

Arias, E., "Linking students to urban livestock producers in Mexico City", *Urban Agricultural Magazine,* 1(2). Leusden, the Netherlands, RUAF, 2000.

Atta-Krah, A.N., "Fodder trees and shrubs in tropical Africa: importance, availability and patterns of utilization", In: *Integration of livestock with crops in response to increasing population pressure on available resources,* (Editors: T R Preston, M Rosales and H Osorio), CTA: Ede, Netherlands, 1991.

Barry, T.N. and Blaney, B.R., "Secondary compounds of forages", In: *The nutrition of herbivores,* (Editors: Hacker, J.B. and Ternouth, J.H.). Academic Press. Australia, 1987.

Beck-Chenoweth, H., "Free-range poultry production & marketing: A guide to raising, processing, and marketing premium quality chicken, turkey & eggs", Back Forty Books, Creola, Ohio, 1997.

Bogenfürst, F., "The current state and future of incubation in waterfowl", World's Poultry Science Association, Preliminary Proceedings, 10th European Symposium on Waterfowl, Halle, Germany, 1995.

Brown, Edward, *Poultry Breeding & Production,* London, 1929.

Buckels, C.G., "Midwest poultry producers share success stories", *Small Farm Today;* May 2003.

Bussi, B., *Disease as a constraint on productivity in extensively managed poultry,* University of Edinburgh, Edinburgh, 1987.

Calnek, B.W., Barnes, H.J, Beard, C.W., Reid, W.M. & Yoderg H.W.Jr., *Diseases of Poultry,* Ninth Edition, Iowa State University Press, Ames, U.S.A., 1991.

Chabeuf, N., "Disease prevention in smallholder village poultry production in Africa", *Proceedings International Seminar on Smallholder Rural Poultry Production,* 9-13 October, 1990.

Cicero, K., "Homes on the range: Portable poultry pens are proliferating", *The New Farm,* May/June 1995.

Cunnane, S.C. and L. U. Thompson, editors—*Flaxseed in Human Nutrition,* AOCS Press, Champaign, 1995.

De Boer, J., "Technological and socioeconomic changes: including urbanization, as they impact upon animal production in Asia", *Animal Production and Rural Development: Proceedings of the Sixth AAAP Animal Science Congress,* Vol I. Bangkok, 1992.

De Haan, C., Steinfeld, H. & Blackburn, H., *Livestock and the environment: finding a balance,* Fressingfield, Eye, Suffolk, UK, WREN Media, United Nations, United States Agency for International Development and the World Bank, 1996.

De Nardi, R., "Results of two years of influenza surveillance in wild birds and backyard flocks of Northern Italy", *Western Poultry Disease Conference: Proceedings of the Western Poultry Disease Conference,* 2007.

FAO, "Livestock's contribution to the protection of the environment", by C. Dalibard, *World Animal Review*, 1995.

Gardin, Y., "Vaccination against H5N1 highly pathogenic avian influenza: some questions to be addressed", *Western Poultry Disease Conference: Proceedings of the Western Poultry Disease Conference*, 2007.

Habitat, *An urbanising world: global report on human settlements*, United Nations Centre for Human Settlements, HABITAT, New York, Oxford University Press, 1996.

Hays, F. A., *Poultry breeding applied*, Poultry-Dairy Publishing Co., Mount Morris, Ill., 1888.

Holden, S., Ashley, S. & Bazeley, P., *Improving the delivery of animal health services in developing countries: a literature review*, Somerset, UK, Livestock in Development, 1996.

Humphrey, J., "The classification of world livestock systems", *A study prepared for the Animal Production and Health Division*, FAO, AGA/MISC/80/3, 1980.

Jahnke, H.E., "Livestock production systems and livestock development in tropical Africa", Kiel, Germany, 1982.

Johnson G, and Wingfield W. W., *The Poultry Book*, London, 1853.

Klober, K., "Sustainable poultry for pasture", *Small Farm Today;* 15(2):21-22, Apr/May 1998.

Lee, C., "Development of live attenuated vaccine against avian influenza", *Western Poultry Disease Conference: Proceedings of the Western Poultry Disease Conference*, 2007.

Leeson, S. & Summers, J.D., *Commercial poultry nutrition*, University Books, Guelph, Canada, 1991.

Leong, V. Y., "Retrospective review of avian influenza H5N1 control in Hong Kong 1997-2007", *Western Poultry Disease Conference: Proceedings of the Western Poultry Disease Conference*, 2007.

McCorkle, C.M. & Mathias, E., "Paraveterinary healthcare programmes: a global overview", *In* K.-H. Zessin, ed. *Livestock production and diseases in the tropics: livestock production and human welfare*, Proceedings of the VII International Conference of Institutions of Tropical Veterinary Medicine, Vol. II, p. 544-549, Berlin, 25-29 September 1994.

Merritt, E.S. & Clarridge, R.W.E., "The effect of length of holding on the hatchability of goose eggs", *Poultry Science*, 1959.

Naidu, A.S., editor, *Natural Food Antimicrobial Systems*, 2000.

Norton, R. A., "The potential weaponization of avian influenza virus - a possible threat to the commercial poultry industry", *Western*

Poultry Disease Conference: Proceedings of the Western Poultry Disease Conference, 2007.

Rosinski A. & Wencek E., "Current problems on waterfowl breeding and production (in Polish)", *Proceedings of Symposium "Poultry-Farm"*, Gdynia, Poland, 1997.

Rosinski A., Rouvier R., Guy G., Rousselot-Pailley D., & Bielinski A., "Possibilities of increasing reproductive performance and meat production in geese", *Proceedings, XX World's Poultry Congress,* New Delhi, India, 1996.

Rosinski A., Rouvier R., Wezyk S. & Sellier N., "Reproductive performance on geese kept in different management systems", *Preliminary Proceedings, 10th European Symposium on Waterfowl,* World's Poultry Science Association, 18-25, Halle, Germany.

Ruthenberg, H., *Farming systems in the tropics,* Clarendon Press, Oxford, 1980.

Sim, J.S. and S. Nakai, editors, *Egg Uses and Processing Technologies - New Developments,* CAB International, UK 1994.

Slaybaugh-Mitchell, Tracy L., *Indigenous livestock production: An annotated bibliography of husbandry practices (Bibliographies in technology and social change series),* 1995

Smith, O.B. & Olaluku E.A., "Peri-urban livestock production systems", *Cities Feeding People Report Series,* 24. Québec, Canada, IDRC, 1998.

Sonaiya, E.B., "Poultry husbandry in small rural farms", *Entwicklung und landlicher Raum,* 1990.

Thear, K., *Free-Range Poultry,* Farming Press, 1997.

Traupman, M., "Profitable poultry on pasture", *The New Farm,* May/June 1990.

Weber, T. P. and Stilianakis, N. I., "Ecologic immunology of avian influenza (H5N1) in migratory birds", *Emerging Infectious Diseases,* 2007.

Whyte, R.D., *The use and misuse of trees and shrubs as fodder,* Commonwealth Agricultural Bureau. Publication No.10., 1947.

Willin, E.S., "Relation between egg weight and intensity of growth in geese", *Preliminary Proceedings, 10th European Symposium on waterfowl,* World's Poultry Science Assoacition, Halle, Germany, 1995.

Wilson, T., "Integrating livestock and crops for the sustainable use and development of tropical agricultural systems", AGSP-FAO, 1994.

Wint, W. and D. Bourn, *Livestock distribution and the environment in sub-Saharan Africa,* ERGO, Oxford, 1994.

Index